T0205475

Sustainable Textiles: Production, Processing, Manufacturing & Chemistry

Series Editor

Subramanian Senthilkannan Muthu, Head of Sustainability, SgT and API, Kowloon, Hong Kong

More information about this series at http://www.springer.com/series/16490

Subramanian Senthilkannan Muthu ·
R. Rathinamoorthy

Bacterial Cellulose

Sustainable Material for Textiles

Subramanian Senthilkannan Muthu
SgT Group and API
Hong Kong, Kowloon, Hong Kong

R. Rathinamoorthy
Department of Fashion Technology
PSG College of Technology
Coimbatore, Tamil Nadu, India

ISSN 2662-7108 ISSN 2662-7116 (electronic)
Sustainable Textiles: Production, Processing, Manufacturing & Chemistry
ISBN 978-981-15-9583-7 ISBN 978-981-15-9581-3 (eBook)
https://doi.org/10.1007/978-981-15-9581-3

© The Editor(s) (if applicable) and The Author(s), under exclusive license to Springer Nature Singapore Pte Ltd. 2021
This work is subject to copyright. All rights are solely and exclusively licensed by the Publisher, whether the whole or part of the material is concerned, specifically the rights of translation, reprinting, reuse of illustrations, recitation, broadcasting, reproduction on microfilms or in any other physical way, and transmission or information storage and retrieval, electronic adaptation, computer software, or by similar or dissimilar methodology now known or hereafter developed.
The use of general descriptive names, registered names, trademarks, service marks, etc. in this publication does not imply, even in the absence of a specific statement, that such names are exempt from the relevant protective laws and regulations and therefore free for general use.
The publisher, the authors and the editors are safe to assume that the advice and information in this book are believed to be true and accurate at the date of publication. Neither the publisher nor the authors or the editors give a warranty, expressed or implied, with respect to the material contained herein or for any errors or omissions that may have been made. The publisher remains neutral with regard to jurisdictional claims in published maps and institutional affiliations.

This Springer imprint is published by the registered company Springer Nature Singapore Pte Ltd.
The registered company address is: 152 Beach Road, #21-01/04 Gateway East, Singapore 189721, Singapore

Preface

The textile and fashion industry is one of the most polluting industries in the world. Though several policies are proposed by various governments, we can see a huge gap concerning the implementation. The sustainability of a textile product can be brought in many ways like material, chemicals, production methods and also through designs. However, the prevailing fast fashion concept and more frequent trend change in the market act as a barrier in implementing those concepts. Out of all sustainability issues, the use of unsustainable raw material contributes a large amount of impact on the environment. Hence, the fashion and textile industries are looking for potential alternative sustainable materials for its use. Bacterial cellulose is one such sustainable material, which has suitable characteristics for clothing applications.

In our research, we found numerous interesting aspects of bacterial cellulose. Though quite a lot of research works performed in the biotechnology field, the textile and fashion applications explored very little. We found enormous scope in this area but, at the same time, lack of awareness among the researchers in the textile and fashion domain. Hence, to provide insight and to enhance the understanding of the bacterial cellulose and its properties to the textile and fashion fraternity, the contents are designed. The book is specifically in black and white by focusing on textile and fashion students, faculty and researchers in mind. The technical aspects of the production and analysis were detailed in such a way that the non-biotechnological personals can also effectively comprehend the production methods and other details related to microorganisms.

Chapter 1 of the book elucidates the impact of the fast fashion raw materials on the various environment domains including material depletion, unethical wage, water consumption, landfills, use of pesticides, etc. The chapter also addresses the recent sustainable alternatives with huge potential in the textile and fashion fields including mycelium and bacterial cellulose.

From the second chapter onwards, the main emphasis is given to the bacterial cellulose. Chapter 2 details the identification and progression of bacterial cellulose over time. Besides, it details various bacterial strains that are capable of producing cellulose. The various production methods, influencing parameters like carbon, nitrogen source, fermentation time, pH and other media components on cellulose development are provided to enrich the readers' knowledge.

Chapter 3 of the book specifically focuses on the characteristics of the bacterial cellulose. It strengthens various properties like morphology, chemical, physical, thermal and moisture management properties of the bacterial cellulose. Further, huge stress is given in the latter part of the chapter, which particulars the post-treatment process of bacterial cellulose. The post-treatment steps like scouring (purification), bleaching, colouration and functionalisation with different chemicals are very vital in the aspects of textile and fashion application. This content will open new avenues in the reader's mind to use bacterial cellulose in clothing applications.

The last chapter of the book recapitulates the application potential in the area of clothing, accessories, leather alternatives and home textile products. The final part of the last chapter details various other interrelated applications like wound dressing, filtration, composites, etc., wherein the textile fabrics are commonly used. While the previous Chaps. 1 and 2 are dealing with fundamentals of production, the last two chapters will provide a lot of potential information to readers in the technical aspect to nurture their future research.

Kowloon, Hong Kong Dr. Subramanian Senthilkannan Muthu
Coimbatore, India Dr. R. Rathinamoorthy

Acknowledgements

At the very first place, the authors would like to thank **God Almighty** for giving them the might, knowledge, ability and chance to undertake this book and to persist and complete it satisfactorily.

Authors would like to know the students of PSG College of Technology, Department of Fashion Technology, **Ms. T. Kiruba and Ms. S. Raja Balasaraswathi**, for their tremendous contribution in the area of bacterial cellulose development and characterisation.

Additionally, authors want to express their gratitude to their family members for their great support and encouragement to bring out this book in a short span of time.

Contents

About the Authors

Dr. Subramanian Senthilkannan Muthu currently works for SgT Group as Head of Sustainability, and is based out of Hong Kong. He earned his Ph.D. from The Hong Kong Polytechnic University, and is a renowned expert in the areas of Environmental Sustainability in Textiles and Clothing Supply Chain, Product Life Cycle Assessment (LCA) and Product Carbon Footprint Assessment (PCF) in various industrial sectors. He has five years of industrial experience in textile manufacturing, research and development and textile testing and seven years of experience in life cycle assessment (LCA), carbon and ecological footprints assessment of various consumer products. He has published more than 75 research publications, written numerous book chapters and authored/edited multiple books in the areas of Carbon Footprint, Recycling, Environmental Assessment and Environmental Sustainability.

Dr. R. Rathinamoorthy, working as an Associate Professor in the Department of Fashion Technology, PSG College of Technology, Coimbatore, India from 2009. He had completed his Ph.D. on "Medical Textiles" in 2016. Recently he received the "Young Achiever Award" for the year 2019 by the Institute of Engineers India (IEI), coimbatore chapter. In 2017, He also received a national-level "Young Engineer award for the year 2016–17" in Textile Engineering domain, by Institute of Engineers India (IEI), Kolkata, West Bengal, India. He is having a google H index of 11 and a Scopus H index of 7 with more than 400 citations. He had published 19 national and 53 international research articles in various refereed and non-refereed journals. He had presented three international conference papers and two national seminars. He authored 5 technical books in the area of Apparel and fashion technology field and 19 book chapters with various international publishers like Woodhead Publisher, Springer Verlag, Springer Nature, Springer Singapore, Taylor & Francis and Elsevier publishers to his credit. In 2011, he had sanctioned a research project from University Grants Commission for the tune of Rs.10.15 Lakhs and successfully completed it in 2013. He also sanctioned a minor project to the worth of rupees 0.15 lakhs under PSG STEP, an Initiative by DST, NSTEDB, Government of India.

He has eleven years of teaching experience in the area of Textile and Fashion discipline. He also has one year industrial experience as an industrial engineer. His research interest is on sustainable material for textile and fashion industries. He is currently working on bacterial cellulose and other biomaterials to identify their potential in the fashion industry as a sustainable alternative to existing conventional materials.

Chapter 1
Sustainability and Fashion

Issues with Raw Material

1.1 Introduction

For the past few years, the most fascinating topic which is getting greater acceleration in the global fashion market is 'fast fashion'. Actually, the term 'fast fashion' refers to the clothing that quickly moves from catwalk to stores with the purpose of mimicking the current luxury trends. It is an approach to make the fashion trends available for the normal consumers quickly and cheaply. This is the concept by which normal people can also be fashionable as the celebrity and high fashion models with trendy clothes which are being sold at affordable price. To understand better, it refers the fast production of cheap clothes inspired by great fashion houses where the production is made faster so that the time taken to reach the customer will get reduced. Fast fashion is also called 'McFashion' which has the meaning of fast and convenient fashion which is made possible by the fast production and affordable price. This fast fashion system is being fuelled mainly by cheaper clothing, increase in the hungriness for fashionable clothing and increase in the purchasing power of the people. Cheaper clothing attracts middle-class people to go with the trend. Most of the fashion companies are adopting fast fashion strategy by replicating streetwear and fashion week trends as they appear in real time. They are managed to sell trends at greater speed at affordable price. The increase in the clothes consumption and a decrease in the trend cycle duration make the consumer to prefer low-quality and trendy clothes. From the consumers' point of view, the three main reasons that make the prevalence of fast fashion are that it is cheaper, trendy and disposable. The concept is executed efficiently by promoting the continuous cycle of planned production, distribution, disposal and replacement of fashion products. The main elements of fast fashion are reported in Fig. 1.1 as mentioned in the literature [1].

The quick response strategy is one of the common supply chain strategies and extensively used in firms that produce fast fashion material. It integrates the customer requirements and production capacity to meet the market needs on time. This process integrates all parts of the supply chain and assures quality. Further, it shares the

© The Author(s), under exclusive license to Springer Nature Singapore Pte Ltd. 2021

S. S. Muthu and R. Rathinamoorthy, *Bacterial Cellulose*, Sustainable Textiles: Production, Processing, Manufacturing & Chemistry,
https://doi.org/10.1007/978-981-15-9581-3_1

Fig. 1.1 Main elements of fast fashion [1]

information with vendors and retailers through online mode, and so, no work in progress delays at any stage of the manufacturing process. The design is one of the important restrictions for the shorter production cycle. Hence, the second element either generalises the design or delays the customer's buying behaviour until the design in the retail store for sale. The term agility represents the flexibility of the manufacturing firm to develop different styles and mass quantities as the market requires, usually in a short span of time. In order to achieve this effect, the supply chain connection should be more market-sensitive to achieve customer requirements whenever the demand arises. This further allows the vendors and manufacturers to be virtually connected and update the real-time data. Hence, the unnecessary work in progress avoided and complete integrated network connection make the production possible at any place. Finally, the assortment, it is a retailing practice which makes the customer not to get bored among the existing collection. As the fast fashion collection launched in a shorter time, the dynamic assortment of production makes the customer's shopping experience more interesting and invites them for frequent shopping [1]. This enables the manufacturers and brands to deliver the product on time to the market. A recent research report indicated that the environmental impact from the apparel manufacturing companies will increase by 80% in 2025 if the same level of per capita consumption follows [2]. Now, the fast fashion giants are offering products on a weekly basis, and they call it as microseason, over 3–4 seasons per year.

This ultimately increases the production requirement from the manufacturers and significantly increased the environmental impact [3]. The following section details the major influencing factors of the fast fashion system.

1.2 Impacts of Fast Fashion

1.2.1 Overproduction

The 'fast fashion' trend leads to the situation of making more quantity with less quality. The growing demand leads to increased production. With a big demand for new clothes, if there is some error in forecasting, a company will lose its profit if it underestimates its demand. Hence, with the low production cost, the companies overestimate their future sales. This ends in overproduction [4]. In the last 15 years, clothing production is doubled, and it is clear that this increase in the production level is due to the fast fashion phenomenon [5]. In a year, nearly 150 million garments are produced in the global fashion industry which roughly means about 20 items per person [4]. Overproduction which is a part of fast fashion has the disadvantages of material exploitation as well as environmental hazards.

1.2.2 Material Exploitation

Textile and fashion industries rely more on non-renewable resources for its raw materials like oils for the production of synthetic fibres, pesticides and fertilisers for the cultivation of cotton, chemicals for the processing and finishing of textile materials. The demand for cotton and polyester gets increased as they play a major role in fast fashion. Cotton and polyester account for nearly 24.4% and 51.5%, respectively, of global fibre production [6]. The fashion industry uses 1.5 trillion litres of freshwaters for processing, whereas in the world around 750 million people do not have access to fresh drinking water.

Cotton being a natural fibre that requires large cultivation lands and polyester is a synthetic material produced from a non-renewable resource, and their overconsumption affects the ecological balance. Moreover, the overproduction of clothing to match the consumer needs also has the negative outcome that it is sometimes not sold completely. With the urge of updating the stores with new collections, unsold clothes are sent to landfill as waste. Nearly 30% of the clothes produced are not sold and end up as waste [7]. The fashion industry produces nearly 92 tons of solid waste each year which utilises 98 million tons of natural resources. And some brands like Burberry have the practice of burning the unsold garments worth up to 37 million dollars to avoid the discount sales [4]. As most of these are made of synthetic materials, they result in affecting environmental sustainability. An estimate states that by

2030, if the trend goes on like this, the fashion industry will face a reduction in profit by 52 billion dollars which may due to the scarcity in the raw materials and workers [4]. For example, every year, 98 million tons of crude petroleum is used for polyester manufacturing alone. This is approximately 1% of the oil produced worldwide. If the apparel industry produced in the same phase for the future, it will reach 300 million tonnes of oil by 2050 and responsible for 26% of the CO_2 emission. Currently, it represents 10% of the emission. In comparison with cotton, polyester emits three times more carbon dioxide for every T-shirt produced [8].

1.2.3 Reduced Quality

The clothing products made by fast fashion brands are of tremendously low quality owing to the deployment of cheap and low-quality materials and cheap labour, and eventually, fast fashion products will degrade very soon. They are intended to serve a lower functional life as these products are given no care to the quality. The main intention of fast fashion is to make products that will not last for long; thereby, one can purchase more products. With few exceptions to be noted of course, but the majority of fast fashion is perceived to be discarded after a few uses or wears. The concept of fast fashion relies on a production model that depends on a rapid replenishment cycle of damn low-quality garments at very cheaper price points [9]. Due to these, new styles are delivered to store on a monthly or even weekly basis [10] which will encourage overconsumption of garments [11].

1.2.4 Unethical Practices

While working towards the goal of reducing manufacturing cost, there happens a violation of human rights in terms of an unsafe working environment, child labour and forced labour. Many industries are compromising the working environment for cutting down the price. This affects workers' safety and health. Workers are made to face a dangerous environment while working with low-quality toxic chemicals and poorly conditioned machines. Since most of the brands outsource labour-intensive work, they are not paying more attention to the working conditions of the factories. Even though some of them do audits on such conditions, their effectiveness is not at that level. This fails in attaining the societal aspect of sustainability [12]. Forced labours and child labours are hired in fashion industries so that they can reduce the labour cost. Particularly, for cotton cultivation, most of the man powers used are child labours. The next problem is wages. Under the pressure of creating more jobs, the government has kept the minimum wages as too low, which is not sufficient. And many of the companies continued to use cheap paid labours to supply low-cost products [12]. In Bangladesh, a worker of the textile industry earns only 197 dollars per month which are one-fourth of the amount required for the fulfilment of the basic needs of a worker [12]. The factories developing low-cost products mainly operate in developing countries, where 85% of the fashion industry workers are women, where labours are asked to work for a longer duration without essential workplace requirements. The workers are exposed to more than 3000 different chemicals in various stages of apparel production. The recent Rana Plaza accident in Bangladesh was one of the best examples of an unsafe workplace. The accident killed around 1134 workers, mostly women. Similarly, another incident happened in India's textile hub, Tamil Nadu, where more than 100 workers were drugged to work for a longer time without any fatigue during work time in the fast fashion industry. The workers were not aware of the drug, and later, it produced health issues like fibroids, depression, miscarriages and anxiety [13]. More recently, The New York Times reported that the Fashion Nova's production chain is under the surveillance of the Federal Labour Department for wage theft. The report mentioned that the popular brand subcontracted the vendor factories and owed a large sum of money up to USD 3.8 million as a back wage. In some factories, the labours were paid as little as USD 2.77 per hour. Fifty Fashion Nova factories are under scrutiny in this issue. The federal officials told the newspaper that the scrutiny was to make sure that the company is paying a fair wage to the workers [14]. These are the wake-up call for the world to do some scrutiny on the unethical practices performed in fast fashion industries.

1.2.5 Environmental and Health Hazards

The detrimental effects of different life cycle phases of textile products starting from raw material and encompassed through manufacturing, distribution and consumer use and disposal are well discussed in the literature [15–21]. By means of producing cheap quality products in a short time, fast fashion adds further to the severity of the environmental and health hazards of textile and fashion products, which can be understood from the wealth of the information available from the literature. Fast fashion makes more new products in the collection and so increases the impact on the environment. From the transportation stage to disposal stage, overconsumption adds more impact to environment in terms of energy, water and carbon footprints, human toxicity, eutrophication, etc. [15, 17, 21].

1.3 Raw Materials—The Major Cause for Unsustainability

The textile and fashion industry is the reason for 5.4% of the world's pollution and is considered as the fifth most unsustainable industry preceded by electricity and heat, agriculture, road transportation, and oil and gas production [22]. As discussed in the previous sections, the fashion industry has a considerable impact on sustainability starting from the raw materials to finished goods. Whenever the impact of the fashion industry on the environment is taken into the study, the major role in polluting the environment is being played by the raw materials. The raw materials used in the industry have a serious impact on the environment than the effects of processing. A huge amount of resources is being exploited in the process of obtaining raw materials for the fashion industry. The production of cotton requires a large amount of cultivation land and water. And the production of synthetic fibres relies on non-renewable resources, and their extraction process involves the usage of high-energy machinery [6]. Moreover, the effect of synthetic materials on the environment after production is also hazardous. Various problems with the raw materials which are currently in use are discussed here.

1.3.1 Impact of Cotton

Cotton is one of the most popularly used fibres in the apparel and fashion industry because of its ultimate comfort properties. Cotton, being a natural fibre, is more often considered as a sustainable, environmental-friendly material due to its biodegradability. Yet, the production and processing of cotton fibres have considerable risks on the environment as well as social sustainability.

Pesticides and Insecticides: Cotton is a crop prone to pests. 2.5% of the world's agricultural land is used for cotton cultivation, and the cultivation of cotton uses 16% of the total insecticides used for agriculture [23]. Roughly, for growing one pound of cotton, it requires 1/3 lb of insecticides [24]. This much usage of insecticides has adverse effects on soil fertility. Moreover, this has the risk of affecting the water bodies as well, when the pesticides are washed out of soil [25]. Another problem with pesticides is that it can destroy some beneficial fauna while fighting against pests [26]. The World Health Organisation has considered cotton as the 'dirtiest crop' as half of the chemicals used in the cotton cultivation have a serious effect on human and environment [25]. Nitrogen-based synthetic fertilisers are being used in cotton cultivation. One of the insecticides 'Aldicarb' which is used in cotton cultivation has the potential to kill a human when even one drop of it enters into the human body through the skin [24]. In Uzbekistan, which accounts for nearly 4–5% of the world's cotton cultivation, the groundwater up to 150 m depth got polluted because of the hazardous pesticides. And this unsafe drinking water results in the poor health of around 85% of the population. A report says that nearly 20,000 people are dying of cancer and miscarriage annually due to the effect of chemicals sprayed on cotton. The breast milk of women living near the intensive cotton cultivation region is detected with a high concentration of pesticides [27].

Overconsumption of Water: Cotton is also denoted as 'thirsty crop' as the water requirement for cotton cultivation is enormous [28]. Cotton consumes nearly 2.6% of global water use [29]. It requires nearly 7000 to 29,000 L of water for the production of 1 kg of cotton fibre [28], while, compared with synthetic textiles, the production of polyester consumes less than 0.1% of the water required for cotton cultivation [27]. A report says that water consumed for India's cotton exports in the year 2013 would be enough to provide 1.24 billion people with 100 L of water every day for one year

[30]. The Aral Sea, which is once the largest lake in the world, is virtually gone now due to cotton cultivation. Not only the water depletion, but the region also suffers from almost 80% of the world's throat cancer due to the hazardous pesticides and insecticides used in cotton production [26]. The inefficiency in water management in developing countries is another reason for huge water consumption. For instance, the US cotton utilises 8000 L of water for producing 1 kg of cotton, whereas, in India, it requires up to 10,000 L of water [30]. Though it is not a hugely significant value, cotton contributes nearly 0.3–1.0% of global carbon emissions [31].

Unethical Practices: 99% of the world's cotton cultivation is made in developing countries where the labour, health and safety regulations are poor. This often leads to negative effects such as child labour, forced labour and unhealthy working environment [32]. The workers in the cultivation field are not provided with the proper safety guidelines and basic safety needs like gloves, respirator glasses while spraying pesticides which can create an adverse effect on them. The cotton cultivation is a labour-intensive task and that high labour demand leads to child labours and forced labours [26]. Children starting from the age of five are being exposed to work on the cotton field with little or no pay. In Uzbekistan, around 1.5–2.0 million children are forced to work in the cotton industry during harvesting season, and this action is a government-sanctioned one [33]. Children who were made to work in ginning industries without proper protective clothing and masks are often suffered from respiratory problems due to air contaminated with cotton dust in the industry [26].

1.3.2 Impact of Synthetic Textiles

Synthetic materials account for 63% of total textile production. While considering the synthetic textiles, the dominating fibres are polyester (55%), nylon (5%) and acrylic (2%) [33]. The synthetic fibres are held in the top place of usage because of their excellent flexible characteristics in terms of both manufacturer and consumer sides. The demand for synthetic materials especially polyester keeps on increasing and its ultimate lower-cost fuels fast fashion [34]. Though their production processes are much simpler than the processing of natural fibres in terms of cultivation land and water requirement, the risks with these materials are relatively very high [35].

Non-biodegradability: The first and foremost problem with synthetic textiles is its non-biodegradability. It will take up to 200 years for the polyester to decompose [36]. About 72% of the clothing we use are made of synthetic fibres [37]. In USA, textile wastes account for around 6% of municipal solid waste every year. Though the waste includes natural and synthetic fibres, synthetic materials, due to their non-biodegradability, end in landfills and cause huge long-term environmental impact [38].

Non-renewable resource depletion: The production of synthetic fibres relies more on non-renewable resources [36]. Polyester, being the most dominating synthetic fibre, it should be noted that it is produced with the by-product of petroleum [35]. It is produced out of the chemical reaction which involves coal, petroleum, air and water [34]. For the production of polyester fibres alone, nearly 70 million barrels of oil are being consumed per year [36]. In the year of 2015, synthetic fibre production consumed more than 330 million barrels of oil [34].

Toxic and energy-intensive production process: Since the production often relies on oil, the extraction of oil itself results in causing pollution in the form of discarded waste and oil spilling [35]. The production process of polyester from petroleum is a toxic process that has the potential to cause hormonal disruption and breast cancer [35]. The production process is also energy-intensive and has the chance of producing solid, liquid or gas by-products which can cause undesirable impacts on the environment [39]. The production process of nylon involves the generation of nitrous oxide. It is a known fact that nitrous oxide is five times more dangerous than carbon dioxide which is affecting the ozone layer [35]. The production of polyester involves 63% more energy than that of cotton production [26]. For the production of one kilogram of polyester, 125 MJ of energy is consumed [34]. Though the water consumption is not high as cotton, polyester production also consumes a huge amount of water for the cooling down process as the production process involves high heat generation. Moreover, the dyes used for polyester fabrics are disperse dyes which are made up of complex molecular structures. Disperse dyes are insoluble in water and are not readily decomposed and have the potential to create environmental impact [40].

1.3.3 Impact of Leather

Leather, a classic luxury material, is made out of the skin of dead animals. The demand for leather products never fades because of their durability and luxury look. Leather is not a sustainable material as the processing results in a huge impact on environmental sustainability. They are contributing more to global warming and pollution. Higg Materials Sustainability Index by Sustainable Apparel Coalition has given a rating as 159 for leather's impact [41]. An estimate by the Food and Agriculture Organization estimates that 3.8 billion bovine animals are used every year for leather production. Though the leathers are often made from the by-product of the meat industry, sometimes animals are slaughtered for the luxurious and exotic leather varieties [42].

Environmental impacts: The process of converting animal skin into leather involves the usage of a huge amount of energy and chemicals [43]. The environmental and social concerns arise with the tanning process. Tanning is the most important and basic process in leather manufacturing where the collagen in the animal skin is stabilised in order to stop the degradation of the skin [43]. Chromium, having a potential impact on the environment, is the most commonly used chemical in the tanning process. Almost, 80% of the tanning process is done using chromium [44]. Tanneries produce a huge volume of organic and inorganic wastes that have the potential of polluting soil, atmosphere, water and air. Out of raw hides, only 15% is converted into leather, whereas the rest 85% is generated as waste from the processing. The wastes include chemicals, animal hair, keratin and flesh [42]. For the tanning process, around 50–60 L of water are used per kilogram of hide, and they are discharged into the water bodies along with the added chemicals after the process [45]. The wastewater from the tanneries is composed of 250 mg/L of chromium concentration and 500 mg/L of sulphide concentration [46]. Every day, around 22,000 L of waste from tanneries is dumped into the river in Bangladesh [44]. When these wastes are dumped into water bodies, they can affect the living beings in the water bodies. They are potential enough to damage fish gills and cause respiratory problems, infections, infertility and birth defects [47].

Unsafe environment for Workers: Workers working in the leather processing units are not often provided with safety measures which can cause skin, eye and respiratory diseases. Around 16 million people in the world are at the risk because of the exposure to the hazardous chromium salts which is used in the tanneries [44]. Chromium salts are potential enough to cause swelling, inflammation, dermatitis and eczema when it comes to contact with the skin [46]. When the chromium gets absorbed into the skin, they can cause erosive ulceration often called 'chrome holes' [47]. Most of the people working in the processing of leather are suffering from Leukemia. A report by the 'Centers for Disease Control and Prevention' showed that the occurrence of leukemia among the people residing near a tannery in Kentucky is five times more than the average of the USA [43]. In India, a range of 2.2–38% of workers working in the tanneries is being affected by Asthma [45]. Dimethylformamide (DMFA), the solvent used in leather production, has potential impacts on the workers who are exposed to it. A report says that women exposed to DMFA while working in the leather industry for a period of 1–5 years are suffering negative effects on pregnancy and childbirth [45].

1.3.4 Impact of Regenerated Cellulosic Fibres

Viscose Rayon, Modal Rayon and Lyocell are commonly known regenerative fibres. These are man-made cellulosic fibers. Regenerated fibres account for 6–7% of total fibre production [48].

Deforestation: The regenerated fibres are produced mainly from wood pulp particularly from the Eucalyptus tree. This can create a negative impact in terms of deforestation. 70 million trees are cut down every year for the production of clothing. A report says that nearly 30% of the rayon and viscose fabrics are produced by cutting down of trees in more endangered and ancient forests [44].

Toxic Chemicals in the processing: The spinning process of viscose fibres involves the usage of various chemicals including caustic soda, carbon disulphide, sulphuric acid, sodium sulphate and zinc sulphate. When compared with cotton, the requirement of energy is more in the regenerated cellulose fibre production. Though cotton requires high water consumption, it requires 8–52% less energy than regenerated cellulose production [48].

1.3.5 Impact of Wool

Right from breeding sheep to mothproofing garments, every production step from raw material stage to manufacturing phases of the life cycle, there are many devastating impacts on the environment (land, air and water) [49]. According to the ranking of 'Pulse of the Fashion Industry' report, the production of sheep's wool was rated as more polluting—for cradle-to-gate environmental impact per kilogram of material—than that of acrylic, polyester, spandex and rayon fibres [50]. One of the major issues pertaining to wool is greenhouse gas emissions from sheep. Sheep are found to release enormous amounts of methane gas (CH_4) into the atmosphere [51]. Methane has 28 times more global warming potential than carbon-di-oxide as per the fourth assessment report of greenhouse gas protocol [52]. Additionally, manure generated by farmed animals has significantly contributed to the rapid rise in atmospheric greenhouse gases over the last 250 years [51].

1.4 Sustainable Alternatives

In order to overcome the negative impacts of traditional raw materials, efforts are made to find sustainable alternative materials. There are many sustainable alternatives tried in the textiles and clothing sector, namely soybean fibre, pineapple fibre, citrus fibre, banana fibre, peace silk, organic wool, milkweed fibre, fibres extracted from milk, recycled fibres, etc. Some of the sustainable alternatives which are not previously discussed in detail are listed below:

1.4.1 Citrus Fibre

Citrus fibre is an innovative sustainable alternative for the existing unsustainable textile raw materials. The fabrics are produced by using the by-products of citrus juice. Over the past five decades, a steady rise in the food processing industries has been seen. This also leads to the creation of a huge amount of non-edible wastes out of the process. Though the impact of that waste on the environment is not significant, it is considered as unsustainable practice as this leads to wasting our valuable natural resources. The great effort to use that non-edible waste from the food processing industry results in such a sustainable textile material. Fabrics made out of citrus fruit are patented by orange fibre. It is a cellulose-based material. The citrus fruit wastes often known as 'pastazzo' which are left after the squeezing of juice are processed to extract citrus cellulose. The extracted citrus cellulose is apt for spinning to produce yarn. Those yarns are used in the making of high-quality, luxurious and sustainable fashion products [53]. This can be a sustainable alternative for the regenerated cellulosic fibres made out of wood pulps. They are claiming that the textile materials made out of 100% citrus material have features like soft and silky hand, and they can be made opaque or shiny as per the demands [53].

1.4.2 Pineapple Fibre

Pineapple fibre is a growing sustainable alternative of leather. The fibres are produced from the by-products of pineapple harvests without the need of any extra resources [26]. The fibres are considered to be more sustainable as there are no toxic chemicals used in the processing of leaf to form a fibre. Generally, the fibres are extracted from the leaves through scraping followed by the retting process. Scraping is done either by hand or by machines. Hand scraping utilises broken porcelain plates, whereas, in decorticating machines, revolving blades will do the process. After scraping, the fibres will be thoroughly washed with water and dried. Then, the process of retting is done for loosening of fibres by removing various cementing tissue components. Retting is a microorganism-based process [54]. Hence, the production process of pineapple fibres does not involve hazardous chemicals. Another thing to be noted with pineapple fibres is that the cultivation of pineapple does not require a huge amount of pesticides and insecticides, unlike cotton cultivation. This makes pineapple fibres more sustainable than traditional raw materials. Yet it is not 100% sustainable as polyurethane coating is done to make the material more apt for fashion applications. This can be avoided in the future by finding a suitable alternative to the polyurethane coating. The pineapple textiles can also be made without PUV coating where the durability will be less than that of traditional material. In such cases, the pineapple material will be a sustainable material [55].

1.4.3 Mycelium Leather

As learnt earlier that the leather making process is environmentally devastating, there are a few attempts to produce leather using vegan inputs and one such is mycelium leather. Mycelium, which is the root structure of fungi like a mushroom that just grows by the absorption of nutrients from the environment [56], is presented as a solution to replace the resource-intensive conventional leather and mycelium is produced by depending upon nature's best tools [57]. This sustainable, versatile and animal-free option of leather can grow quickly from mycelium by using a carbon-negative process [57]. The idea of reducing the use of the non-renewable resource, which is being a part of a sustainable development strategy, paves the way towards sustainable bio-based material. Bio-based materials are those which can be grown sustainably and fully renewable [9]. Mycelium is one of such sustainable bio-based materials, having flexible properties so that they can be used in a variety of applications.

Mycelium is the vegetative part of the fungi, and it is the root structure of the fungi [58]. 'Hyphae', the tubular filaments, are the fundamental unit of the filamentous fungi [59], and mycelium has the porous structure [58] which is the network of randomly oriented interlinking hyphae [59]. Hyphae are fine fibrous material whose diameter ranges from 1 to 30 μm, and the length varies from microns to metres. The length of the hyphae varies with the species, substrate and the growing environment [58]. Mycelium can be grown by inoculating the species on a suitable substrate or medium that can provide the essential nutrients for growth. The most capable fungi taxa for producing mycelium are Ascomycota and Basidiomycota [60]. Usually, the nutrients needed for mycelium growth are carbon, nitrogen, vitamins and minerals [59].

Mycelium materials can be produced as both composites and pure mycelium [59]. The pure mycelium can be cultivated in the liquid culture under static or agitated conditions. While making composites, the mycelium is grown into the substrate materials to interlock with those substrates to form bulk materials [59]. The production of mycelium involves little energy, and they composed completely of renewable materials [60]. Moreover, the agricultural wastes such as rice and wheat straw, cottonseed hull [61], white maize, red sorghum, guinea corn, yellow maize, millet, wheat seeds [62] and hemp pith [63] can be used as a potential substrate for mycelium growth [59]. And also, mycelium is considered as 'recycler organism' as it can disassemble the large organic molecules into simple ones [64]. Huge waste from mechanical processing of jute in industries which are having no other use than fuel for a boiler can have the potential to be a substrate to grow mushroom [65]. These aspects of mycelium make it a sustainable material.

Mycelium is a natural polymeric composite material that is mainly composed of natural polymers such as chitin, cellulose and proteins [66]. The properties of mycelium produced often depend on the medium or substrate on which it is grown and the type of the species used. They absorb nutrients from their surrounding, further their growth pattern will change with the surrounding and the growing atmosphere [67]. Fine-tuning of the medium components can result in flexible properties [68]. Further, the materials are found to have good mechanical [69], fire resistance [70] and acoustic properties [71]. This advantage of mycelium material makes it a material with a wide variety of applications.

The idea of using fungal mycelium in the textile industry is due to its resemblance to lignocellulosic fibres. In the textile and fashion industries, mycelium materials are considered as a potential alternative for leather materials [68]. Bolt threads have developed Mylo™, which is a leather-like material that is made out of mycelium. Their idea of using mycelium as a substitute for leather aims to reduce the environmental impact of leather. And another advantage of mycelium is that they can be grown in weeks, whereas for producing leather it takes more time [72]. When it comes to making garments, mycelium is considered for making simple parts; that is, instead of producing fabrics and cutting them, mycelium can be grown in the shape of the garment parts [67]. Currently, experiments are being done to develop mycelium materials that can imitate the velvet fabrics in terms of shine and tactile by growing mycelium on a base cloth. Further, the experiments are being made to get the fluid nature of the velvet fabrics in mycelium materials [64]. Moreover, mycelium composites are found as a suitable sustainable alternative for footwear components like shoe soles. The attributes of the material such as support to the wearer's foot and cushioning against the rough ground make it apt for the replacement of unsustainable shoe soles [73]. Mycelium films are also used in the wound dressing application, as the fibrous structure and the cellulosic content of mycelium are effective in entrapping the drugs [74]. However, researches are being made to find suitable growing techniques to make use of mycelium widely in the textile industry.

1.4.4 Bacterial Cellulose

Bio-based materials and bio-compatible materials are in need to make the existing unsustainable textile and fashion business to be sustainable. A lot of fashion business giants, designers and researchers are seeking sustainable alternatives to the fashion sector to address the unsustainability issues of the existing raw materials. One of such materials is bacterial cellulose or biocellulose, which is again cellulose, obtained through the bacterial fermentation process. It is eco-friendly, safe to the human body and renewable raw material [75]. It is also identified as one of the sustainable alternatives in the leather and footwear sector too [76]. Also known as microbial cellulose, bacterial cellulose is a cellulosic material, which is derived from bacteria, which can produce cellulose. It is an organic compound with the formula $(C_6H_{10}O_5)n$ produced from some of the types of bacteria [77]. One of the salient features of biocellulose is that it is produced without the impurities present in plant-derived cellulosic products (namely lignin and hemicellulose) [78]. Biocellulose possesses unique features that make as a potential sustainable source, which are biocompatibility, biodegradability, ultra-fine molecular structure, desired mechanical strength, non-toxic nature and its purity and neutrality.

This entire book is intended to discuss in-depth and detail about bacterial cellulose—production, manufacturing, properties, characteristics in terms of textiles and fashion perspective and finally various applications of bacterial cellulose.

References

1. Aggour A, Moussaid A, Abou El Hassan A (2018) Fast fashion assessment tool: a case study of a moroccan apparel supply chain. Int J Sup Chain Manage 7(5):213–220
2. Remy N, Speelman E, Swartz S (2017) Style that's sustainable: a new fast-fashion formula. http://www.mckinsey.com/business-functions/sustainability-and-resource-productivity/our-insights/style-thats-sustainable-a-new-fast-fashion-formula. Accessed 16 Jul 2019

3. Modi D (2013) Upcycling fabric waste in design studio, thesis submitted to National Institute of Fashion Technology, Mumbai. http://14.139.111.26/jspui/bitstream/1/71/1/upcycling%20fabric%20waste%20in%20design%20studio.pdf. Accessed on 16 Jul 2020
4. Seven forms of sustainable fashion (2019) https://www.greenstrategy.se/sustainable-fashion/seven-forms-of-sustainable-fashion/. Accessed 30 Jul 2020
5. Joy A, Pena C (2017) Sustainability and the fashion Industry, conceptualizing nature and traceability. In: Book sustainability in fashion: a cradle to upcycle approach. https://doi.org/10.1007/978-3-319-51253-2_3
6. Slater Keith (2003) Environmental impact of polyester and polyamide textiles. Woodhead Publishing, UK
7. Sphera (2020) Life cycle assessment. https://sphera.com/insights/what-is-a-life-cycle-assessment/. Accessed 30 Jul 2020
8. Wagner L (2020) The environmental impact of the fast fashion industry. https://sanvt.com/journal/environmental-impact-of-fast-fashion-infographic/. Accessed 30 Jul 2020
9. Levy M, Weitz BA (2008) Retailing management, 7th edn. McGraw-Hill Irwin, Boston
10. Palomo-Lovinski N, Hahn K (2014) Fashion design industry impressions of current sustainable practices. Fashion Pract J Des Creative Process Fashion Ind 6(1):87–106
11. Rissanen T (2013) Zero-waste fashion design: a study at the intersection of cloth, fashion design and pattern cutting. PhD Thesis, University of Technology, Sydney
12. Negative Impacts of Fast Fashion from around the world to Turkey. Design a new model for territorial valorisation and increased awareness (2019)
13. Phadtare A (2020) Fast fashion, loose ethics: the human and environmental cost of cheap clothing and what we can do about it. https://www.thelovepost.global/protection/articles/fast-fashion-loose-ethics-human-and-environmental-cost-cheap-clothing-and-what. Accessed 30 Jul 2020
14. Mlotek H (2019) Fast fashion at the end of the world. https://newrepublic.com/article/156065/fast-fashion-end-world. Accessed 30 Jul 2020
15. Muthu SS (2014) Assessing the environmental impacts of textiles & the clothing supply chain. Woodhead Publishing, UK
16. Muthu SS (ed) (2015) The sustainable apparel production handbook. CRC Press, USA
17. Muthu SS (ed) (2015) Handbook of life cycle assessment in textiles and clothing. Elsevier Publishing, UK
18. Muthu SS (Ed) (2017) Sustainability in textile industry. Springer, Berlin
19. Muthu SS (Ed) (2017) Textiles and clothing sustainability: sustainability in textiles and clothing sector & will clothing be sustainable?
20. Muthu SS (ed) (2017) Sustainable fibers and textiles: a complete guide. Elsevier Publications, UK
21. Muthu SS (2020) Assessing the environmental impacts of textiles & the clothing supply chain. Elsevier Publications, UK. (Revision 2, UK, 2020)
22. Willow F (2018) Fashion is not the second highest polluting industry, here are the real numbers. https://ethicalunicorn.com/2018/02/01/fashion-is-not-the-second-highest-polluting-industry-here-are-the-real-numbers/. Accessed 5 Aug 2020
23. The risks of cotton farming. https://organiccotton.org/oc/Cotton-general/Impact-of-cotton/Risk-of-cotton-farming.php. Accessed 5 Aug 2020
24. What's so bad about cotton? (2010). https://business-ethics.com/2010/08/07/1438-the-bad-side-of-cotton/. Accessed 5 Aug 2020
25. Edwards S (2014) Textile review: the environmental impacts of cotton. https://www.tortoiseandladygrey.com/2014/09/01/textile-review-the-environmental-impacts-of-cotton/. Accessed 5 Aug 2020
26. Kalliala EM, Nousiainen P (1999) Life cycle assessment environmental profile of cotton and polyester-cotton fabrics. AUTEX Res J 1(1):8–20
27. Blackburn RS (2009) Sustainable textiles life cycle and environmental impact. Woodhead Publishing, UK

28. Safaya S, Zhang G, Mathews R (2016) Toward sustainable water use in the cotton supply chain. https://waterfootprint.org/media/downloads/Assessm_water_footprint_cotton_India.pdf. Accessed 5 Aug 2020
29. Leahy S (2015) World water day: the cost of cotton in water-challenged India. https://www.theguardian.com/sustainable-business/2015/mar/20/cost-cotton-water-challenged-india-world-water-day. Accessed 5 Aug 2020
30. Chapagain AK, Hoekstra AY, Savenije HHG, Gautam R (2006) The water footprint of cotton consumption: an assessment of the impact of worldwide consumption of cotton products on the water resources in the cotton producing countries. Ecol Econ 60:186–203
31. Our Fiber Eco-Review (2020). https://www.sustainyourstyle.org/en/fiber-ecoreview. Accessed 5 Aug 2020
32. Child labour in cotton: a briefing/International Labour Office (2016) Fundamental principles and rights at work branch (Fundamentals). ILO, Geneva
33. Young S (2019) The real cost of your clothes: these are the fabrics with the best and worst environmental impact. https://www.independent.co.uk/life-style/fashion/fabrics-environment-fast-fashion-eco-friendly-pollution-waste-polyester-cotton-fur-recycle-a8963921.html. Accessed 5 Aug 2020
34. Fibre briefing: polyester (2018). https://www.commonobjective.co/article/fibre-briefing-polyester. Accessed 5 Aug 2020
35. The Dangers of Synthetic Fibers and Fabrics on the Environment (2018). https://superegoworld.com/blogs/the-world/the-dangers-of-synthetic-fibers-and-fabrics-on-the-environment#:~:text=Synthetic%20materials%20which%20are%20by,less%20harmful%20to%20the%20society. Accessed 5 Aug 2020
36. What is circular fashion? (2018). https://www.commonobjective.co/article/what-is-circular-fashion. Accessed 5 Aug 2020
37. Synthetic fibres used in 72% clothing items can sit in landfills for 200 years (2019). https://www.sustainablefashion.earth/type/water/synthetic-fibres-used-in-72-clothing-items-can-sit-in-landfills-for-200-years/. Accessed 6 Aug 2020
38. Morgan Elston L (2019) Synthetic clothing contributes to plastic waste in water. https://news.law.fordham.edu/elr/2019/10/27/synthetic-clothing-contributes-to-plastic-waste-in-water/. Accessed 6 Aug 2020
39. Slater K (2008) Environmental impact polyester and polyamide textiles. Polyesters and polyamides, pp 171–199. https://doi.org/10.1533/9781845694609.1.171
40. Uren A (2018) Material guide: how sustainable is polyester? https://goodonyou.eco/material-guide-polyester-2/. Accessed 6 Aug 2020
41. Fibre Briefing: Leather (2018). https://www.commonobjective.co/article/fibre-briefing-leather. Accessed 6 Aug 2020
42. Edwards S (2016) The environmental impacts of leather. https://www.tortoiseandladygrey.com/2016/05/02/environmental-impacts-leather-fashion/. Accessed 6 Aug 2020
43. Environmental Hazards of Leather. https://www.peta.org/issues/animals-used-for-clothing/leather-industry/leather-environmental-hazards/. Accessed 6 Aug 2020
44. Syed Madiha, Saleem Taimur, Shuja-ur-Rehman Muhammed Asif, Iqbal Faisal Javed, Khan Muhammed Bilal Salman, Sadiq Kamran (2010) Effects of leather industry on health and recommendations for improving the situation in Pakistan. Arch Environ Occup Health 65(3):163–172. https://doi.org/10.1080/19338241003730895
45. Tarantola A (2014) How leather is slowly killing the people and places that make it. https://gizmodo.com/how-leather-is-slowly-killing-the-people-and-places-tha-1572678618. Accessed 6 Aug 2020
46. Green insights: newsletter on eco-labelling and eco-friendly products, leather and environment 12(1), April–June 2017
47. Bowyer JL, McFarland A, Pepke E, Groot H, Erickson G, Henderson C, Jacobs M, Fernholz K, Dovetail Partners, Inc. (November 11, 2019) An examination of environmental impacts of clothing manufacture, purchase, use, and disposal
48. Sohel R, Subramani P, Shama P, Raul F (2014) Regenerated cellulosic fibers and their implications on sustainability. https://doi.org/10.1007/978-981-287-065-0_8

49. Environmental Hazards of Wool (2020). https://www.peta.org/issues/animals-used-for-clothing/wool-industry/wool-environmental-hazards/#:~:text=Environmental%20Hazards%20of%20Wool&text=At%20every%20stage%20of%20production,land%2C%20air%2C%20and%20water.&text=Manure%20generated%20from%20livestock%20has,over%20the%20last%20250%20years. Accessed 6 Aug 2020
50. Pulse of the fashion industry (2017) http://globalfashionagenda.com/wp-content/uploads/2017/05/Pulse-of-the-Fashion-Industry_2017.pdf. Accessed 6 Aug 2020
51. The environmental impact of wool. https://theecologist.org/2019/mar/12/environmental-impact-wool. Accessed 6 Aug 2020
52. Global warming potential values. https://www.ghgprotocol.org/sites/default/files/ghgp/Global-Warming-Potential-Values%20%28Feb%2016%202016%29_1.pdf. Accessed 6 Aug 2020
53. Sustainable fabrics from citrus juice by-products. http://orangefiber.it/en/. Accessed 6 Aug 2020
54. Jawaid Mohammad, Asim Paridah Mohammad, Tahir Md, Nasir Mohammed (2020) Pineapple leaf fibers—processing, properties and applications. Springer, Singapore
55. Kate (2018) Pineapple leaf fibre: a new sustainable material. https://timetosew.uk/pinatex-a-new-sustainable-material/. Accessed 6 Aug 2020
56. This very realistic fake leather is made from mushrooms, not cows. https://www.fastcompany.com/40562633/this-leather-is-made-from-mushrooms-not-cows. Accessed 6 Aug 2020
57. Mycelium, redefining leather with the power of organic technology. https://fashnerd.com/2016/11/mycelium-redefining-leather-with-the-power-of-organic-technology/. Accessed 6 Aug 2020
58. Islam MR, Tudryn G, Bucinell R, Schadler L, Picu RC (2017) Morphology and mechanics of fungal mycelium. Scientific Reports 7:13070. https://doi.org/10.1038/s41598-017-13295-2
59. Karana E, Blauwhoff D, Hultink E-J, Camere S (2018) When the material grows: a case study on designing (with) mycelium-based materials. Int J Des 12(2):119–136
60. Lelivelt RR (2015) Mechanical possibilities of mycelium materials, Master thesis, Eindhoven University of Technology. https://research.tue.nl/en/studentTheses/the-mechanical-possibilities-of-mycelium-materials. Accessed 6 Aug 2020
61. Juan HE, Cheng C-m, Su D-g, Zhong M-f (2014) Study on the mechanical properties of the latex-mycelium composite. Appl Mech Mater 507:415–420
62. Stanley Herbert O (2010) Effect of substrates of spawn production on mycelial growth of Oystermushroom species. Agric Biol J N Am. https://doi.org/10.5251/abjna.2010.1.5.817.820
63. Ziegler AR, Bajwa SG, Holt GA, McIntyre G, Bajwa DS (2016) Evaluation of physico-mechanical properties of mycelium reinforced green biocomposites made from cellulosic fibers. Am Soc Agric Biol Eng 32(6):931–938
64. Collet C (2017) 'Grow-Made' textiles, EKSIG 2017: alive. Active. Adaptive. In: International conference on experiential knowledge and emerging materials, Delft University of Technology, The Netherlands
65. Samanta KK, Basak S, Chattopadhyay SK (2015) Recycled fibrous and nonfibrous biomass for value-added textile and nontextile applications. In: Muthu S (eds) Environmental implications of recycling and recycled products. Environmental footprints and eco-design of products and processes. Springer, Singapore. https://doi.org/10.1007/978-981-287-643-0_8
66. Haneef M, Ceseracciu L, Canale C, Bayer IS, Heredia Guerrero JA, Athanassiou A (2017) Advanced materials from fungal mycelium: fabrication and tuning of physical properties. Scientific Reports, 7:41292. https://doi.org/10.1038/srep41292
67. Nimkulrat N, Ræbild U, Piper A (2018) Soft landing, cumulus think tank. Publication No 3 of Cumulus International Association of Universities and Colleges in Art, Design and Media. https://www.cumulusassociation.org/wp-content/uploads/2018/04/PDF_SINGLE_cumulus_soft_landing.pdf. Accessed 6 Aug 2020
68. Mazur R (2015) Mechanical properties of sheets comprised of mycelium: a paper engineering perspective. Honors Theses. 68. https://digitalcommons.esf.edu/honors/68
69. Bruscato C, Malvessi E, Brandalise RN, Camassola M (2019) High performance of macro fungi in the production of mycelium-based biofoams using sawdust—sustainable technology for waster education. J Clean Prod 234:225–232. https://doi.org/10.1016/j.jclepro.2019.06.150

70. Jones Mitchell, Bhat Tanmay, Huynh Tien, Kandare Everson, Yuen Richard, Wang Chun H, John Sabu (2018) Waste-derived low-cost mycelium composite construction materials with improved fire safety. Fire Mater 42(7):816–825. https://doi.org/10.1002/fam.2637
71. Pelletiera MG, Holta GA, Wanjuraa JD, Larab AJ, Tapia-Carillob A, McIntyrec G, Bayer E (2016) An evaluation study of pressure-compressed acoustic absorbers grown on agricultural by-products. Ind Crops Prod 95:342–347. https://doi.org/10.1016/j.indcrop.2016.10.042
72. Bolt Technology, Meet Mylo. https://boltthreads.com/technology/mylo/. Accessed 6 Aug 2020
73. Silverman J (2018) Development and testing of mycelium-based composite materials for shoe sole applications. http://udspace.udel.edu/handle/19716/23768. Accessed 6 Aug 2020
74. Khamrai M, Banerjee SL, Kundu PP (2018) A sustainable production method of mycelium biomass using an isolated fungal strain *Phanerochaetechrysosporium* (accession no: KY593186): Its exploitation in wound healing patch formation. Biocatalysis and agricultural biotechnology, vol 16, pp 548–557. https://doi.org/10.1016/j.bcab.2018.09.013
75. Domskiene J, Sederaviciute F, Simonaityte J (2019) Kombucha bacterial cellulose for sustainable fashion. Int J Clothing Sci Technol 31(5):644–652. https://doi.org/10.1108/IJCST-02-2019-0010
76. Rathinamoorthy R, Kiruba T (2020) Bacterial cellulose—a sustainable alternative material for footwear and leather products. In: Muthu S (eds) Leather and footwear sustainability. Textile science and clothing technology. Springer, Singapore. https://doi.org/10.1007/978-981-15-6296-9_5
77. Luis Q (2017) Embracing bacterial cellulose as a catalyst for sustainable fashion. Senior Honors Theses. 711. Available at: https://digitalcommons.liberty.edu/honors/711
78. Cai Z, Kim J (2010) Bacterial cellulose/poly (ethylene glycol) composite: characterization and first evaluation of biocompatibility. Cellulose 17(1):83–91

Chapter 2
Bacterial Cellulose

Raw Materials and Production Methods

2.1 Introduction

Cellulose is one of the most common polysaccharides used in several applications around the world. Though the plants are the major source of cellulose production, few other sources like vegetables, algae, fungi, and few moulds also can able to produce cellulose. The plant cellulose mainly consists of hemicellulose and lignin with the cell wall. Research reports mentioned that pure form of the cellulose can also be produced from a few genera of the bacterial strains, namely *Agrobacterium, Acetobacter, Sarcina* and *Rhizobium* [1, 2]. In general, bacterial-based celluloses are purer in structure than the plant-based cellulose. Though the structure of both plant-based and bacteria-based cellulose is common, the physical and chemical properties of the bacterial cellulose are different in nature. The national academy of science and technology of the Philippines mentioned that the nata de coco, a bacterial fermented cellulose-based food is identified in the eighteenth century itself as a dessert [3]. In 1886, Dr. Brown discovered the bacterial cellulose. In his research on biochemical actions of *Bacterium aceti* and acetic fermentation process, he noted a lucent mass creation on the culture media surface, which he referred to as 'vinegar plant'. In his research finding, he discovered that the *Bacterium aceti* was not strong and it cannot withstand a simple agitation but the researcher noted that the vinegar plant can withstand and can be strong enough for several days. Based on this findings, he identified that both the films, *Bacterium aceti*, and vinegar plant were produced by two different organisms and he recreated the vinegar plant with the same recipe and evaluated its properties. Brown treated it with strong hydrochloric acid and identified that the jelly-like structure is purely made of cellulose and water through his experiments. After the findings of Brown, in 1949, Teodula K. Africa developed first bacterial cellulose, nata de coco as an alternative to the old Philippine Nata de pina [4]. The research reported by Lapuz et al., scientifically confirmed that Nata is made of the gelatinous pellicle made of cellulose generated by bacteria. As of now, nata de coco of Philippines is the only commercialised form of bacterial cellulose

© The Author(s), under exclusive license to Springer Nature Singapore Pte Ltd. 2021

S. S. Muthu and R. Rathinamoorthy, *Bacterial Cellulose*, Sustainable Textiles: Production, Processing, Manufacturing & Chemistry,
https://doi.org/10.1007/978-981-15-9581-3_2

[5]. Later, several research workers performed research works on the properties of the cellulose developed from the nata de coco in the mid of 1980s [6–8]. Several other research works confirmed the formation of cellulose from the bacterial species through various analysis techniques like X-ray technique [9] and electron microscope analysis [10].

The next significant study performed on the bacterial cellulose production from *Acetobacter* sp. is by Schramm and Hestrin. In their first research, they developed a purification and quantification method for the recovery of pure cellulose from the bacterial cellulose. They verified the suitability of their rapid method for both the plant-based cellulose and bacterial cellulose. In their subsequent study, they evaluated the production of cellulose in a glucose-rich medium using *Acetobacter xylinum* and its mechanism [11]. Their findings are crucial and the medium developed by Schramm and Hestrin became the standard medium for the *A. xylinum.* The findings of their study are [12]:

1. The floatation of the culture from the nutrient medium is due to the formation of microbubbles. This is an effect due to the gas generated by the fermentation process.
2. The production of cellulose happens due to the growth and concentration of cells on the surface. They found that oxygen supply is the essential medium for the growth of cells on the surface and cellulose forming enzymes and thus increases the production of cellulose.
3. As the cellulose production is the function of oxygen supply, the higher production of cellulose happens in the top side and the lower part with minimal activity.
4. Their studies also find that the agitated culture of the same medium showed a different pattern and less cellulose production even though the supply of oxygen was high.

These findings were very vital for successive researches that identified the mechanism of cellulose production from the *A. xylinum* species. In the later year, Malcolm brown and co-workers performed a detailed analysis of the cellulose production from the *A. xylinum* species and their results were noted significant for the research community. The researchers used darkfield light microscopy to analyse the production of cellulose from its structure. The production of the cellulose occurs from the outer surface of the bacterial strain, it can produce up to 50 numbers of individual fibres in a row and the fibre production speed also measured by the researchers. They also mentioned that the cell development process does not limit the cellulose production, even after separation both parent and new cells are capable of producing cellulose in the same aspect [13]. In his next research, the researcher attempted to document the cellulose synthesis with optical brightening agent in the medium using an electron micrograph pictures. In his findings, he reported the role of factors affecting the incubation and culture medium on the cellulose pellicles produced [14]. In the successive research, they had reported the first evidence of visualisation of cellulose formation in vitro. In this research, they performed an observation of cellulose synthesis from *A. xylinum* within a short time. With the dark field and electron microscope analysis, they confirmed the formation of cellulose and noted that the diameter is similar to the cellulose produced from in vivo studies [15]. Though several attempts made on this stage to identify the synthesis mechanism of cellulose, the researchers could not able to find the proper mechanism. This was detailed by Dr. Brown Jr in his review [16]. Further to add, several other research works on the cellulose formation and its structure were detailed by Dr. Brown and his team in different periods [17–20].

2.2 Chemical and Physical Structure

Plant cellulose is the basis for all the cellulose structures. It is a linear polymer of D-glucopyranose sugar units connected by β linkages. The plant cellulose possesses a very high degree of polymerisation, and so it has good mechanical properties. Out of its structure, more than 65% is pure crystalline and cannot be accessible by any solvents. The presence of lignin and hemicelluloses in the outer sheath of the polymer acts as a protective coating for the cotton from the degradation [21]. In chemical structure, both the cellulose namely plant and bacterial cellulose possess the same molecular formula. However, their physical properties are drastically varying. The bacterial cellulose is a highly pure form of cellulose with a higher crystallinity percentage. The individual fibre diameter of the bacterial cellulose is a hundred times smaller than the plant cellulose fibre. This physical structure change ultimately increases the tensile and moisture-related properties of the bacterial cellulose (Fig. 2.1).

Bacterial cellulose is very similar to the plant cellulose with respect to its chemical structure. The bacterial cellulose structure also consists of a linear homopolymer of glucose monomers, and they were connected by glycosidic linkage similar to plant cellulose. But at the same time, the researchers mentioned that the degree of polymerization ranges between 2000 and 6000, which is far lower than the plant cellulose. Bacterial cellulose is identified as one of the purest forms of cellulose. The fundamental chemical structure of the bacterial cellulose consists of successive glucose units linked with 180° rotation. The thickness and the quality of the cellulose produced mainly depend upon the type of bacterial species and the medium used in the production. However, in its structure, many cellulose chains interlinked through the hydrogen bonds to form a sheet [21, 22]. Researchers reported that during the production, the cellulose molecule was extended and produced in the form of a ribbon. This process makes the hydroxyl groups in the cellulose project

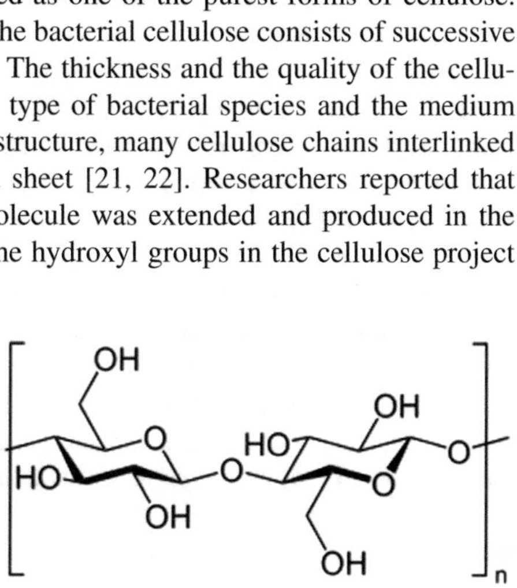

OH
OH
O
HO
O
O
HO
O
O
OH
OH
n

Fig. 2.1 Chemical structure of the cellulose [Reprinted with permission]

outside or outer surface of the structure. The researcher reported this might be the reason for the higher water absorption capacity of the bacterial cellulose. The second effect developed by the stretching creates a direct link between carbon and hydrogen atoms in the molecule, and this creates a stronger hydrophobic structure [23]. The strains produce two different structures of cellulose by nature. The first structure is cellulose I, more crystalline one which is created by the ribbon-like arrangement of fibrils during production. The second structure is cellulose II, and this is a thermodynamically stable amorphous polymer [17]. Yu and Atalla reported that the higher crystalline nature of the cellulose I is associated with its uni-axial arrangement of 1-4 glucan chain with van der Waals force. Similarly, the random arrangement of the 1-4 glucan with hydrogen bond is the main reason for the characteristics of cellulose II in bacterial cellulose [24]. The self-assembling nanofibrous structure is one of the unique physical nature of the bacterial cellulose over the plant cellulose. This physical arrangement with nano-sized fibre provides all the special properties like higher tensile, crystalline nature and water absorption over plant cellulose [25].

2.3 Cellulose Producing Bacteria

The production of cellulose from bacteria is reported from several types of species. In the identified bacterial strains, most of the types namely *Acetobacter, Azotobacter, Rhizobium, Pseudomonas, Salmonella, Agrobacterium, Aerobacter, Achromobacter* and *Alcaligenes* are gram-negative species. The gram-positive species like *Sarcina ventriculi, Salmonella* and *Escherichia* are also capable of producing cellulose. Out of all the bacterial strains identified *A. xylinum* is noted as one of the most promising strains in terms of cellulose production [26]. It is reported that under perfect conditions, the bacteria can convert almost 50% of the carbon source used in the medium to cellulose in the form of a pellicle. In specific, *A. xylinum* is a gram-negative and aerobic strain. The bacterium is in rod shape and produces cellulose by its primary metabolic activity in the structure of interwoven ribbons. One of the advantages of the *A. xylinum* is its versatile cellulose producing nature concerning the various carbon sources. In general, the bacteria can produce cellulose under two different conditions. (i) Static condition and (ii) agitated condition. Under static condition, the bacterial culture along with the culture medium left in the aerated dark place undisturbed till the fermentation happens. In the agitated culture method, the setup is stirred continuously throughout the storage time. In the first case, the cellulose pellicle created as a continuous mat-like structure, and in the latter case, though the cellulose is grown aster, it produces round balls of cellulose.

Research reports identified that the cellulose production can be correlated with the cell concentration, and they noted that it is in a linear correlation with cell concentration [27]. In *A. xylinum*, the cellulose production is mainly produced as a protective mechanism of its structure. The growth of the cellulose pellicle on the surface protects the bacterial cell from the UV light's killing effect. Many research workers observed that the cellulose production by *A. xylinum* can be improved by

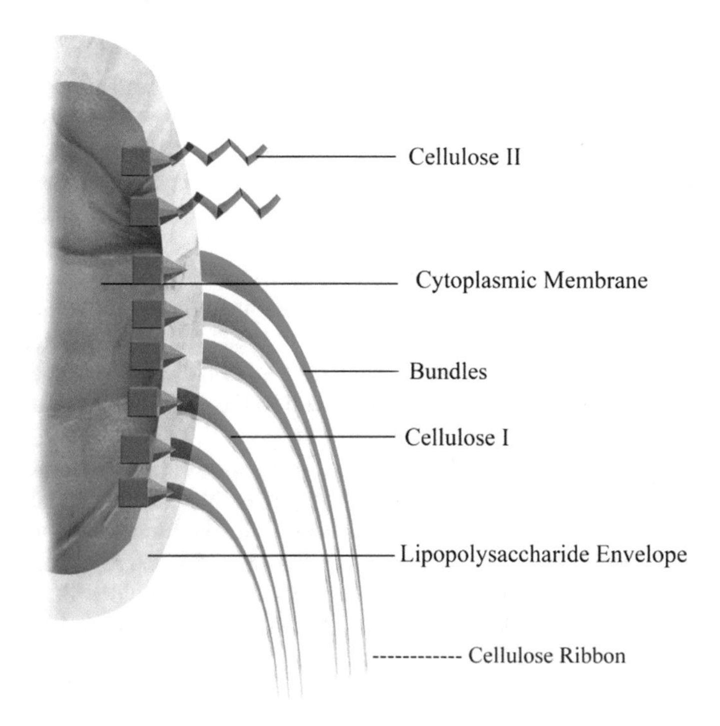

Fig. 2.2 Cellulose production from bacteria [Authors own illustration]

increased oxygen supply. The cellulose produced in the outer cell wall and cytoplasma membrane through enzyme called cellulose synthase [28]. Here, the synthesised cellulose released through the pores on the surface of the cell wall. The produced fibrils are nano-sized individual fibres; however, due to the sequential arrangement of pores the released fibrils align together in the form of ribbons to form crystalline cellulose as shown in Fig. 2.2 [29, 30]. The researchers also confirmed that the cellulose formation occurs only in the air/liquid interface and not on the medium or inside the medium [12, 31].

Agrobacterium tumefaciens is another type of soil bacteria, which readily produce cellulose from glucose. But the cellulose production with this bacteria happens in the formation of floc. Though the cellulose production on the *A. tumefaciens* occurs on the outer cell wall like *A. xylinum,* the structural properties of the cellulose produced noted as different. The research report by Matthysse analysed the cellulose produced by *A. tumefaciens* in detail and characterised its properties [31]. The researcher developed the cellulose from this strain using carrot tissue culture cells and characterised it through enzyme and infrared analysis. Further, the visible formation of cellulose flocs within the bundle of bacterial strains are noted in the scanning electron micrograph images. A detailed analysis of the cellulose produced from *A. tumefaciens* using XRD analysis revealed that the cellulose is mainly made of cellulose I [32].

Rhizobium spp. is another type of bacteria, which synthesises cellulose in the presence of carbon sources. *Rhizobium* spp. produces the cellulose fibril during the growth phase as an extracellular product [33]. In the static production process as similar to *A. xylinum,* it can able to produce cellulose pellicle on the surface. Researchers found that the produced cellulose is Cellulose I type after a series of characterisation. Smit et al. reported the production of cellulose through electron micrograph analysis and reported that the cellulose produced by *Rhizobium* spp. is approximately 5–6 nm in diameter and length varies from 1 to 10 micrometres [34]. Researchers mentioned that the cellulose produced in the subsequent process of cell attachment and confirmed it with the attachment loss ability of the fibril negative bacterial spices [33, 34]. *Sarcina ventriculi* is another bacteria known for cellulose production. In this species, the cellulose production is on the cell wall pockets of the cell and it helps the structural formation of cellulose known as pockets. The structure of the cellulose obtained from these bacteria is different from other species [35]. Other researchers reported that the cell pockets and its cementing material create the cellulose. They identified the cellulose and reported a purification method to isolate the cellulose from the cells [36]. The different structural cellulose production might be attributed to the gram-negative cell structure of this bacteria, and the cellulose identified from the species is noted as cellulose II type [37]. Kazim reported the potentiality of *Pseudomonas* isolated from different sample foods and its cellulose production with the help of HS medium. In this study, they confirmed the cellulose production with FTIR analysis [38]. As similar to *Rhizobiaceae,* the *Gluconacetobacter xylinus* also able to produce cellulose on the surface of the rotten fruits and the reports mentioned that the *G. xylinus* produces cellulose as a function of maintaining an aerobic environment in liquids. In the case of *Salmonella* spp. and *E. coli,* the cellulose production is noted as thin aggregative fimbriae.

Though several species are reported to produce cellulose, *A. xylinum* identified as one of the major strain types that produce more amount of bacterial cellulose in terms of quality and quantity. The superior properties of the bacterial cellulose like mechanical, moisture handling, and other structural aspects are the main reason for its higher potential application than plant cellulose. It is also to be noted that the bacterial cellulose will not replace the plant cellulose in terms of application. The major difficulty noted in this process is limitations associated with the industrial scale or mass production feasibilities.

2.4 Bacterial Cellulose—Production Process

For the production of bacterial cellulose, the most common and fundamental methods used are static and agitated culture methods of production. The different bacterial cellulose production methods are provided in Fig. 2.3.

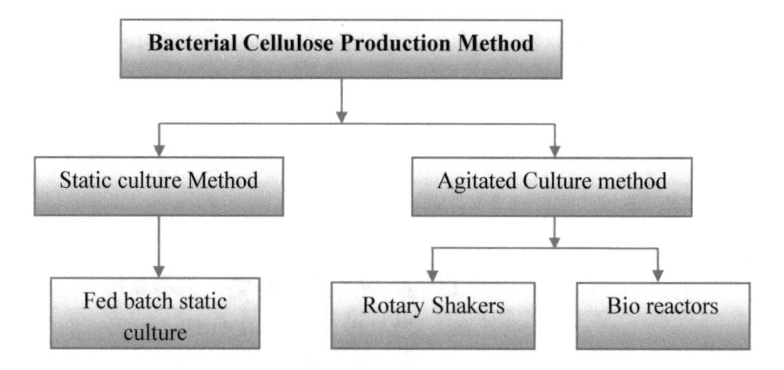

Fig. 2.3 Classification of bacterial cellulose production method [Authors own illustration]

2.4.1 Static Culture Method

In this method, the nutrient medium is prepared as per the bacterial strain requirements. After the inoculation, the medium is kept stationary without disturbing. In the case of cellulose production, to increase production during static conditions, the medium is kept in an aerated place to ensure the proper oxygen supply. Further, few research workers also reported that the absence of light sometimes also increases the production of the cellulose from *Acetobacter xylinum* species. The static culture method is one of the most commonly used methods, and the main disadvantage of this production method is its longer fermentation time. The other disadvantage is that during the fermentation time, the nutrient and oxygen supply is improper for different bacteria present at a different location. This in turn creates a difference in bacterial cellulose production at different places, and thus, it creates an uneven cellulose mat with different thickness at places. Hsieh et al. reported a modified method in the static culture method itself, by feeding nutrient medium intermittently during the cultivation time. In this process after certain days of fermentation, a fresh nutrient medium is fed into the same static culture medium as a second batch. The fed medium is poured directly on top of the newly formed pellicle, during the previous fermentation process. This process can be repeated until several layers of pellicles formed. The researcher reported that this method produces continuous cellulose production of 0.02 g/day even after the 30 days cultivation, whereas in the older cultivation method, it is zero [39].

2.4.2 Agitated Culture Method

In contrast to the static method, the inoculated culture medium is kept in a rotary shaker throughout the fermentation or incubation period, where the continuous agitation of the culture increases the oxygen supply in the liquid medium with the *A. xylinum* bacterial strain. But this method will produce cellulose in the shape of balls.

Hence, the method is suited for higher cellulose production with respect to the time than the static medium.

2.4.3 Bioreactors

As mentioned earlier, the cellulose production is mainly a function of pressure and amount of oxygen supply in the production medium irrespective of the type of medium used [40]. Hestrin and Schramm identified and reported that relatively a large amount of cellulose can be produced with the static production medium and not with the agitated method. However, the agitated culture can produce cellulose quicker than the static culture, due to the improved oxygen supply in the medium [41]. By considering several factors, few of the latest research also suggested that the agitated culture method holds the best potentials for mass production [33, 42]. The culture condition is one of the important parameters that decide the properties of the cellulose produced. Mohite et al. evaluated and optimised the different roles of medium components on the mechanical properties of the cellulose developed. Out of both the methods, the static method is the best method for cellulose production in the form of a sheet than the agitated method. The cellulose production in the form of a sheet is critically preferred due to its wide scope in different fields of applications [43].

Due to the limitation associated with the static culture method like longer culture time and cost involved in cellulose production, various improved reactor-based methods also suggested by different researchers. At the same time, the new methods must also address the main issue of pellet-shaped cellulose production, which is of less application potential and low crystalline in nature. Researchers developed a bioreactor to feed both oxygen and glucose content directly into the cellulose producing cells of the bacteria [44]. In this process, the researcher designed a bioreactor which dispenses the glucose supply as the aerosol into the culture box. The reactor increased the speed of the cellulose production significantly for a longer period than the older methods of production. The other advantage of the bioreactor is the constant increment in the volume of cellulose produced. As the aerosol is dispensing the glucose from the top, the chances of the formation of by-products are limited and lead to high-quality cellulose production. However, the researcher noted the issue of contamination with the developed system, after 40 days, due to unidentified reasons. The researcher reported a denser fibril structure than the conventional method [44]. Shin-Ping Lin et al. developed a semi-continuous production method of bacterial cellulose using a rotating disc bioreactor. In this process, the researcher used a biocomposite material (plastic composite support) made of agriculture waste as a surface of the bioreactor. The researcher preferred this material as the literature supported the usage of this material may increase the production of bacterial cellulose. In this process, the supporting materials are fixed in an agitator screw which is immersed in the medium. The continuous rotation of the shaft motivates the cellulose production. In this method, after five days of incubation, the formed cellulose is removed as a batch, as the less availability of composite surface may reduce the bacterial cellulose after the 5 days of production. Through this reactor, they obtained the highest productivity of 0.24/g/L/day. The research also identified that the use of plastic composite support helped the bacteria attachment on its surface and reduced the re-inoculations in the successive batches. However, the researchers reported a reduced crystalline nature in the cellulose produced under this method. This further influences the stress and strain behaviour of the produced cellulose from this method [45].

In the case of stirrer type bioreactors the suspension of cellulose in the fluid with bacterial cells makes the liquid more viscous, and thus, it restricts the flow of oxygen in the liquid. Chao et al. developed an internal loop airlift bioreactor for bacterial cellulose production for the first time, with higher efficiency than the stirred tank reactor. This reactor increases the dissolved oxygen in the medium and increases the production of cellulose in the medium. Due to the higher oxygen flow, a higher production rate of 0.116 g/lL/h is obtained in this process. These types of bioreactors produce cellulose pellicles with elliptical fibre morphologies [46]. Wu and Li noted that the normal airlift reactor also has a disadvantage of exposing bacteria to the atmosphere and a very small batch of cellulose produced as an outcome. Hence, the researcher proposed a horizontal bioreactor with several net plates inside the reactor for better air circulation. They identified that the use of 10 net plates did not produce any limiting oxygen effect in the fermentation process. The research also noted a correlation between the properties of the produced bacterial cellulose and the amount of aeration (number of net plates used) over the previous literature. The researchers also reported that this was the first attempt to produce the bacterial cellulose in the medium, whereas other researchers only reported the formation of cellulose in between the pellicle and medium interface [47].

Reiniati et al. developed a stirred tank bioreactor for the production of bacterial cellulose. Through continuous stirring, the reactor ensures the proper oxygen supply to the medium. The results of the study reported that the cell density increased with higher agitation rates. This in turn improved the rate of cellulose production. The researchers reported a 0.59 g/L production in the process. They correlated the stirring speed with the production of cellulose and noted that higher agitation speed may yield an increased cellulose production. They mentioned a maximum production of 1.13 g/L yield with a stirring speed of 700 rpm [48]. Figure 2.4 represents the different types of biorectors reported in this section.

Others developed a bioreactor with a spin filter attached to it. In a 2-L jar fermenter reactor, a spin filter is attached with a six-blade propeller at the bottom of the reactor. The spin filter is a perforated wire mesh attached to the agitation shaft of the fermenter. In this process, the researcher increased the oxygen feed and obtained a higher cellulose production of 4.59 g/L of cellulose production upon 140 h of fermentation. This is a comparatively higher production rate than the previous bioreactors. However, the researchers noted a large amount of cell conversion from cel+ to cel− mutants that are not capable of cellulose production [49]. Table 2.1 reports various production methods and their cellulose production capacities in the past decade.

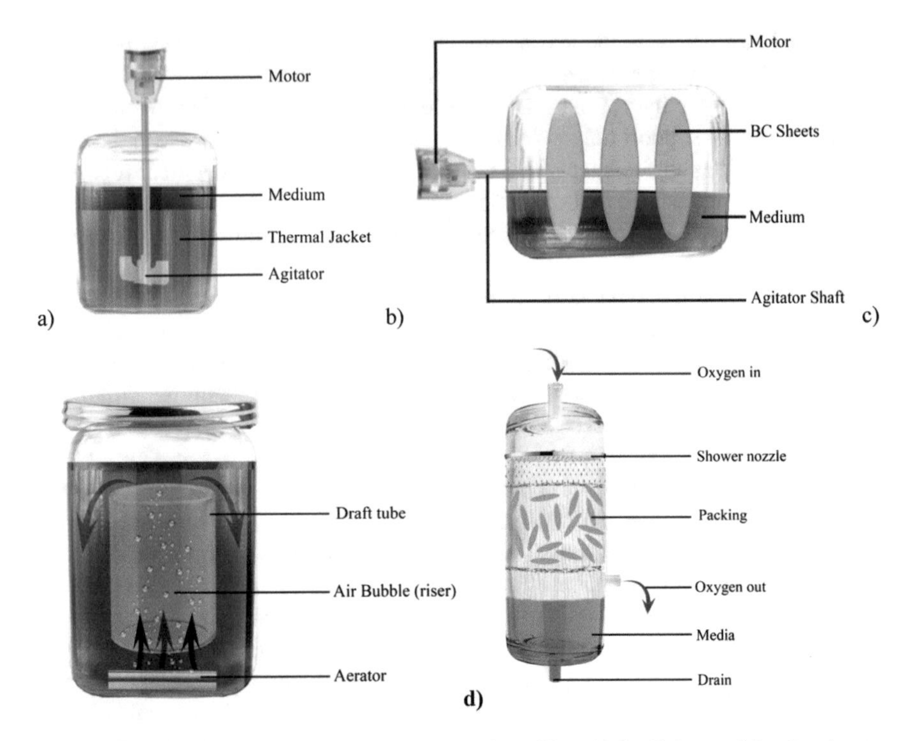

Fig. 2.4 Bioreactor types for the improved production of bacterial cellulose **a** Stirred tank type, **b** Rotating disc type, **c** Air lifter type and **d** Trickling bed type [Authors own illustration]

2.5 Cellulose Production Mechanism

The unique mechanical properties and high purity of the bacterial cellulose are due to the formation of the microfibrillar structure. *A. xylinum* generally produces two different types of cellulose, namely Cellulose I and II. The fundamental mechanism behind the synthesis of cellulose is from the cellulose precursor. The cellulose formation happens by two broad steps, namely (i) polymerisation of glucose units into the $\beta - 1{\rightarrow}4$ glucan chain and (ii) the cellulose chain formation by the crystallization and assembly of the fibrils. More detailed steps involved in the cellulose synthesis from *A. xylinum* can be found elsewhere [65]. The detailed process flow of the mechanism is illustrated in Fig. 2.5.

2.6 Factors Influencing Bacterial Cellulose Production

Culture media has more influence on bacterial cellulose production. Various contents used in the culture media are the factors that have a direct impact on the quality of the cellulose production and also on the quantity of the cellulose. In the meantime,

Table 2.1 Different production methods and cellulose yields as reported in the literature

S. no.	Method used	Cellulose production	Merits	Demerits	References
1.	Bench-scale rotating biofilm contact (RBC) bioreactor	11.65 g/l on 7th day	Higher production	–	[50]
2.	Rotating magnetic fields	–	Higher water holding capacity Potential absorbent material	Design complications and cost	[51]
3.	Silicone tube-based bioreactor	In the form of nanotube for medical implants	Excellent morphology for internal applications	–	[52]
4.	Trickling bed reactor	–	Higher biomass Superior mechanical and water holding capacity	Lower crystalline	[53]
5.	Stirred tank reactor	13 g/L (1.7 fold higher than control)	Increased cell concentration, long-term production	More power required	[54]
6.	Biofilm/polyethersulfone membrane reactor	125 g/m^2, process time 330 h	Cellulose height and weight can be predetermined	Cell fouling	[55]
7.	Biofilm reactor	7.05 g/Lon 5 days incubation	Highly crystalline materials Higher production	Produced cellulose is of cellulose I —β type	[56]
8.	Modified airlift-type bubble column bioreactor	5.6 g/L cellulose after 3 days incubation	Low shear stress and high oxygen transfer rate	–	[57]
9.	Rotary biofilm contactor	5.52 g/L	Aeration rate and number of discs are the important parameter	Culture pH not controlled	[58]
10	Jar fermentor	5.3 g/L	Low cost	–	[59]

(continued)

Table 2.1 (continued)

S. no.	Method used	Cellulose production	Merits	Demerits	References
11.	Bioreactor with rotating cylindrical roller	14.91 g/L	Control on pellicle thickness	No control over cellulose yield	[60]
12.	Rotating disc bioreactor	–	12–20 days production achieved at 3.5 days	Addition of solid particles	[61]
13.	Rotating disc fermentor	3.3 g/L on 7 days cultivation	Slow speed and more surface area developed more cellulose	Young's modulus, water holding capacity noted low	[62]
14	Airlift bioreactor	5.63 g/L in 1 day	Comparable results with Jar fermentor	Environmental influence reduces the mutation	[63]
15	Silicone membrane-based reactor	1.405 g/L after two weeks	5 times higher production than control	The surface of the membrane decides the production	[64]

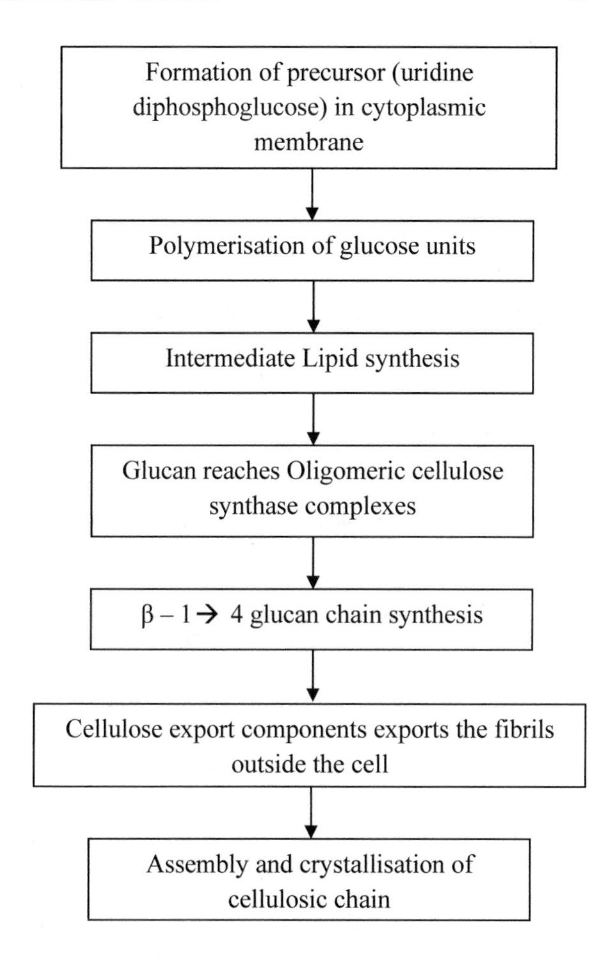

Fig. 2.5 Cellulose synthesis pathway as reported the literatures [14, 15, 66, 67] [Authors own illustration]

it is also important to note that the 30% cost of the bacterial cellulose production is associated with the cost of the culture media used. This restricts the commercial or mass production of bacterial cellulose in the industrial scale. Figure 2.6 represents the probable influencing factors of bacterial cellulose production.

2.6.1 Role of Carbon Source on Bacterial Cellulose Production

Though the standard medium used for the bacterial cellulose is Hestrin and Schramrn (HS) medium, it consists of different synthetic chemical components in it as reported by Schramrn and Hestrin. The carbon source, known as sugar, is one of the essential items for the cellulose synthesis in the bacterial cellulose production. As discussed in the previous section, the use of anyone or multiple types of carbon sources, preferably

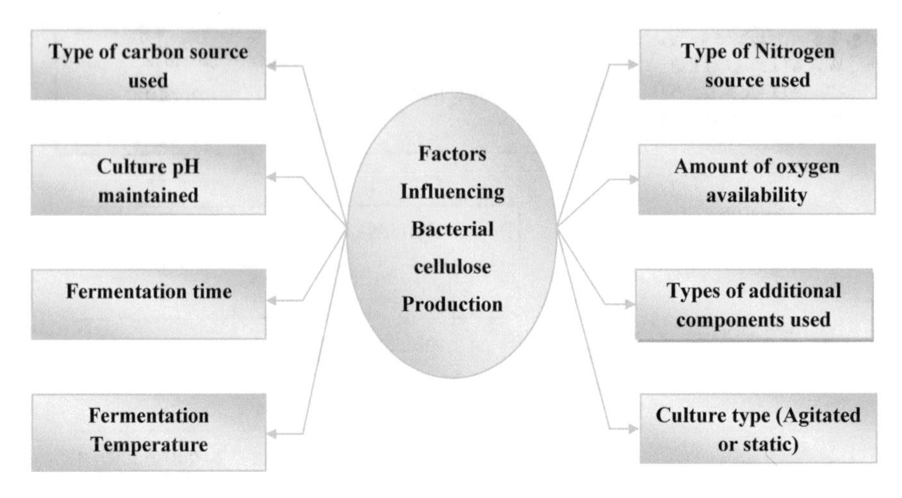

Fig. 2.6 Factors influencing bacterial cellulose production [Authors own illustration]

sugar is essential for cell metabolism and cell growth. Sucrose, glucose and fructose are the main types of sugars commonly used with bacterial cellulose production. The carbon source requirement is different for various bacterial strains. Similarly, every bacterial strain will produce a different amount of bacterial cellulose upon fermentation. The glucose is commonly used for the cellulose production using *A. xylinum* species. The preliminary research performed by Schramrn and Hestrin indicated that the cellulose synthesis in *A. xylinum* using glucose is a function of oxygen supply [1]. However, several research workers performed the possibilities of cellulose production by using sucrose, fructose, maltose, xylose, starch and using glycerol as provided in Table 2.2. The main disadvantage of using glucose as a carbon source is the generation of gluconic acid as a by-product of the process. The development of the acid decreases the pH of the medium and further decreases the cellulose yielding capacity. The second most disadvantage of the most common HS medium and other sugar-based medium are the cost of the medium components.

Many research findings showed that cellulose production primarily depends upon the type of carbon source used. In the meantime, the HS medium or sugar-based medium is costlier to produce the cellulose in industrial (bulk production) aspects. Since then several research workers tried to use various alternative mediums with the food industry and agro-industry wastes as a substitute for the carbon source used in bacterial cellulose production. Out of all the materials discussed, tea is widely accepted and known for its efficient production medium for higher cellulose production. The fundamental works of utilizing the plant extract for the production of cellulose from *A. xylinum* species is done by Brazilian scientists Fontana and their co-workers in 1990. In the subsequent research, they analysed the cellulose producing capability of the few plants including commercial coffee and tea plants. The results of their study reported that in the free sugar media with plant extract supplement

Table 2.2 Research works on glucose-based medium as carbon sources for cellulose production

S. No.	Bacterial strain used	Carbon source used	Cellulose yield	References
1	*Acetobacter xylinum*	Sucrose, glucose, fructose, lactose	5.25–7.38 g/L	[1]
2	*Acetobacter xylinum* ssp. *sucrofermentans* BPR3001A	Fructose with lactate	0.90 g/(L h)	[68]
3	*Acetobacter* sp. A9	Ethanol	15.2 g/L	[69]
4	*A. xylinum* IFO 15606	Xylose	3.0 g/L	[70]
5	*G. hansenii* PJK (KCTC 10505 BP)	Glucose	1.72 g/L	[71]
6	*Acetobacter* sp. V6	Glucose	4.16 g/L	[72]
7	*Acetobacter xylinum* BPR2001	Molasses medium	7.82 g/L	[73]
8	*Acetobacter xylinum* BPR2001	Sucrose	2 g/L	[74]
9	*Gluconacetobacter* sp. RKY5	Glycerol	5.63 g/L	[75]
10	*Gluconacetobacter xylinus*	Corn steep liquor/sucrose	–	[76]
11	*Acetobacter xylinum* ATCC 10245.	Sucrose, fructose, glucose	90, 75 and 77% of cellulose compared to the HS medium	[77]
12	*Gluconacetobacter xylinus*	Mannitol	2.64 g/L	[78]
13	*Gluconacetobacter hansenii* CGMCC1671	Glucose	279.579 mg/g	[79]
14	Gluconacetobacter intermedius	Glycerol	3.4 g/L	[80]
15	*A. xylinum* ATCC 10245	Mannitol	11.11 g/L	[81]
16	*Acetobacter xylinus*	Glycerol	4.95% (v/v) of glycerol added	[82]
17	*Komagataeibacter xylinus* B-12068	Glucose and glycerol	2.4 and 3.3 g/L/day	[83]

showed a better cellulose growth. The tea infusion was observed with higher cellulose production (35.3 g% for 100 ml of liquid and static culture) for the longer fermentation time for more than 7 days. In a shorter time of less than 5 days, coffee found to be effective (14 g%) [84].

The subsequent analysis revealed that all the tested plants had xanthine related contents in its infusion, and this is a bit higher in the case of tea infusion. This suggests the presence of xanthine activates the cellulogenic complex from *A. xylinum* and increases the cellulose production. It is reported that the presence of caffeine, theobromines are the major contributor to cellulose production [84]. In the subsequent year, Reiss (1994) evaluated the fermentation activity of 'tea fungus' (symbiosis of osmophilic yeasts and acetic acid bacteria) under various sugar content in terms of ethanol, lactic acid production along with pH change. Their findings revealed that sucrose had a higher fermentation rate compared to the lactose in terms of higher ethanol production and a tremendous reduction in the media pH. The type of sugar and concentration had very less influence on the fermentation of tea fungus [85]. It is important to discuss the 'kombucha' while we talk about the tea fermentation of tea fungus or *A. xylinum*. Kombucha is a beverage consumed across the globe as a refreshing drink. The beverage is first used in east Asia for the medicinal purpose, and it is originated from China in 220 BC. The culture of kombucha is made of acetic acid bacteria and osmophilic yeast to provide a symbiotic growth. The beverage is prepared by placing the tea fungus (it is not a mushroom, named as the cellulose float's identical look) in a sugared tea infusion after a period of fermentation. The drink can be prepared with any kind of tea leaves (white, green and black) with sugar and can be used as a refreshing beverage. The constituents of kombucha after fermentation were already detailed by other researchers [86]. In this drink, the floating cellulose pellicle will be removed and used for the further fermentation, whereas the medium is used for drinking. Hence, several research works developed bacterial cellulose using the kombucha culture also due to its convenient processing and less hustle. The main advantage of kombucha usage is the higher production of cellulose in the required shape and size.

In recent times, much higher attention is provided on the kombucha-based bacterial cellulose production for several applications due to its simple production steps. Wen Zhang et al. produced bacterial cellulose using the strains isolated from the kombucha culture. Their findings revealed that the cellulose produced using those strains had higher crystallinity, mechanical strength and water holding capacity [87]. Changlai Zhu et al. developed bacterial cellulose directly from the kombucha starter culture. The researchers evaluated the biocompatibility of the developed bacterial cellulose for nerve cells and tissues through in-vivo and in-vitro studies. They also characterised the developed cellulose and reported the effectiveness of the implanted cellulose in rats after 1, 3 and 6 weeks post-implantation. The results were positive and researchers found no significant difference in the blood flow. It is concluded that the biocompatibility of the kombucha-based cellulose is excellent, and they did not exert any hematological and histological toxic effects [88]. Sederavičiūtė et al. reported an environmental-friendly post-treatment for the cellulose developed from the kombucha. The researchers reported a crystallinity of 47% after all the washing and neutralising treatment [89]. Domskiene et al., developed kombucha-based cellulose and evaluated its application in the fashion industry. The researcher developed a static cultured bacterial cellulose film and characterised its property. The researcher evaluated the tensile and mechanical characteristics of the bacterial cellulose after drying it at different temperatures. The researcher concluded that upon drying, the material loses its flexibility and getting stiff. However, they mentioned that the development of the bacterial cellulose similar to the clothing is possible if we can control the thickness of the material after drying. The researcher also suggested opting for different carbon sources as the sucrose or white sugar used in the process is high in cost [90]. The details mentioned above are a small representation of multi-disciplinary attempts by different researchers from the kombucha-based bacterial cellulose. This information is simply a few recent studies nevertheless several researchers performed in the last decade.

Though the tea medium is commonly used for bacterial cellulose, the use of sucrose for a certain percentage is mandatory. The use of tea will not eliminate the complete use of sugar. Hence, concerning the cost-effective production method, several other options to replace the sugar content are also attempted by several researchers. The following are some of the commonly identified alternative carbon sources instead of sugars in the bacterial cellulose production.

2.6.1.1 Grape Medium

Usha rani et al. used grape juice as a medium for bacterial cellulose production. They purchased grape from the local market and pasteurised with 5% sugar. The inoculated *Gluconacetobacter* sp. is used for the bacterial cellulose production process. The bacterial cellulose yield of 7.47 g/L is noted after two weeks of fermentation but at the same time, the yield of the control (HS) is noted as 1.76 g/L. The researcher further characterised the developed bacterial cellulose in terms of its physical and moisture-related properties. The material had very good tensile and elongation properties despite the poor oxygen transmission properties. By considering the thermal and oxygen transfer rate ability, the researchers suggested the developed material as an alternative material for food packaging [91].

2.6.1.2 Rice Wastewater

Rohaeti et al., utilised the house hold rice wastewater, as a potential source for the bacterial cellulose production. The rice wastewater contains starch, protein, minerals and vitamin B, which can be used as a source of nutrients for *A. xylinum* to synthesise cellulose film. Hence, it can be used as both carbon and nitrogen source, the researcher used this along with some sucrose to develop the bacterial cellulose. The results reported a yield of 7.6 g/L of cellulose. The characterisation results reported a higher crystalline (73%), mechanically and thermally stable cellulose through the research [92].

2.6.1.3 Coconut Milk

Coconut milk is identified as one of the nutrient products with Na^+, NO_3^-, NH_4^+ and SO_4 contents. Hence, it is expected to have a significant influence on the yield of bacterial cellulose. Based on the nutrient content, coconut milk is used as a medium along with sucrose for the cultivation of cellulose using *A. xylinum* strain. The results indicated that the use of coconut milk significantly increased the quality and quantity of the bacterial cellulose produced. The experiment yielded a quantity of 43.91 g/L of bacterial cellulose, and this was noted 1.2 times higher than the normal medium without coconut milk [93].

2.6.1.4 Distillery Effluent

In an attempt, researchers used distillery effluent as a medium for bacterial cellulose production by using *Gluconacetobacter oboediens*. The researcher preferred the effluent due to its higher nutrient content. They developed three sets of experiments, one with crude distillery effluent, another set with the addition of sucrose and corn steep liquor as additional nutrient along with the crude effluent and the third set is by adjusting the pH of the second set. After cultivation, the bacteria yield a quantity of 8.5 g/L with crude effluent without adjusting the pH. The second set of experiment with additional nutrient provides further improved yield of 10.8 g/L. However, in the third set, where the pH of the medium adjusted by adding 1.0 N of HCl, in which no growth noted. The researcher inferred that the alteration of pH might have developed some by-products from the effluent and inhibited the growth of the cellulose. However, no specific analysis is performed to confirm the statement. Based on the findings, researchers suggested the use of distillery effluent will be a zero-cost raw material and a suitable alternative material for existing carbon sources [94].

2.6.1.5 Industrial Wastes

Industrial wastes from oil seed-based biodiesel industries and waste streams from confectionery industries are noted as a potential source of nutrients for bacterial cellulose. Using *Komagataeibacter sucrofermentans* strain, researchers produced bacterial cellulose by keeping the crude glycerol and commercial sucrose as a benchmark. In the experiment, it is noted that the cellulose production was improved 4 times than the control experiment. The sunflower meal hydrolysate with glycerol produced 13.3 g/L of bacterial cellulose, whereas the raw glycerol produced 3.2 g/L cellulose and sucrose produced 4.9 g/L cellulose. In the case of flour-rich hydrolysates produced from the confectionery industry used medium, a bacterial cellulose yield of 13 g/L is noted. After evaluating the properties, the researchers proposed by-product streams from the biodiesel industry and waste streams from confectionery industries can be used as a potential source for bacterial cellulose production [95].

2.6.1.6 Waste Glycerol

Ho Jin Yang et al. evaluated the potential application of crude glycerol for the bacterial cellulose production using *G. xylinus* and compared it with pure glycerol production. In this research, the glycerol used is a by-product of the biodiesel manufacturing process. The findings of the research revealed that pure glycerol produced 7.32 g/L of bacterial cellulose and crude glycerol produced a lower amount of cellulose (6.95 g/L) compared to the pure glycerol after 11-day cultivation. The researcher also reported a suitable synthetic medium for the crude glycerol-based medium. By using the crude glycerol, the researchers obtained a maximum of 0.99 g/L cellulose from *G. xylinus* [96].

2.6.1.7 Fruit Juices

Sasithorn Kongruang analysed the production capability of agro-industry wastes like coconut and pineapple juices. The researcher mentioned that the fruit juices are rich in carbohydrates, proteins and trace elements. Three bacterial strains, namely *A. xylinum* TISTR 998, *A. xylinum* TISTR 975 and *A. xylinum* TISTR 893, were used against the selected mediums like coconut and pineapple juice wastes from agro-industries. Along with other media content, the researcher fermented the juices with each strain in a 5-litre container for two weeks. They had reported a yield of 553.33 g/L for *A. xylinum* TISTR 998, 453.33 g/L and 243.33 g/L, respectively, for TISTR 893, TISTR 975 with coconut juice. Similarly, in the case of pineapple juice, a yield of 576.66, 546.66 and 520 g/L is obtained for the above-mentioned strains in the same order. The researchers reported a higher bacterial cellulose production with coconut juice than pineapple and a maximum yield of cellulose noted with *A. xylinum* TISTR 998 in both the cases. The researcher reported that the developed bacterial cellulose is of food-grade and can be used for multiple applications like food, paper and textile industries, without requiring additional steps of decolorization and purification due to its aesthetic and significant mechanical properties [97].

Kurosumi et al. evaluated the possibilities of developing bacterial cellulose using fruit juices of orange, pineapple, apple, Japanese pear and grape. In this research, the fruit pulps were crushed and extracted juices were added with the nitrogen sources used in HS medium. By using *A. xylinum*, the cellulose is developed and production ability was measured. The researcher prepared three mediums with fruit juices like fruit juice adjusted pH 6+ nitrogen source in HS medium (I), fruit juice adjusted pH 6 (II) and sugar reagents + nitrogen source in HS medium (III). Out of all the three selected medium, the recipe I produced bacterial cellulose of 0.48 g. The other medium recipes II and III produced 0.04 and 0.13 g, respectively. The researchers identified out of all, orange was able to produce more cellulose (0.65 g/100 ml). The addition of nitrogen source increased the cellulose production significantly. The researchers also noted that the chemical contents in orange promoting the cellulose production in contrast to other fruits. Hence, they suggested fruits which supports or motivates cellulose production is the best one to use as a nutrient [98].

2.6.1.8 Rotten Fruits and Milk Whey

Rotten fruits from juice and food processing industries and milk whey wastes from the diary industries are used as an alternative potential medium for bacterial cellulose production. The milk whey is one of the serious pollutants for rivers than the sewage effluent due to its ability to create higher oxygen demand in the rivers. Hence, the researchers tried to utilise the wastes as a carbon source for bacterial cellulose production. For the fermentation, the researchers used the *G. xylinus* strain by inoculating in HS medium. The rotten fruits and milk whey are measured for its carbon to nitrogen content ratio and based on the ratio, six culture media (by mixing rotten fruit and milk whey and as an individual) were developed including a control media.

Upon cultivation for 96 h, the results revealed that the medium with rotten fruit alone produced double the time of cellulose than the HS medium control. The cellulose yields of the different medium are, for HS medium—30.8 g/L, milk whey—5.9 g/L, rotten fruits—60.2 g/L, the mixture of rotten fruits and milk whey with different carbon to nitrogen ratio produced a yield of 40–58.2 g/L. Similarly, a maximum amount of crystallinity is noted for the bacterial cellulose produced from rotten fruits as same as HS medium. The researchers reported that this production method was economical, pollution-free and more sustainable [99].

2.6.1.9 Potato Peel Waste

Potato peel waste from the cooking was used as a potential alternative medium for bacterial cellulose production. In this research, the potato peel hydrolysates were prepared as it increases the fermentable sugars in the starch-based substances in the waste. The researcher reported a higher bacterial cellulose yield with the diluted acid hydrolysates than the undiluted acid hydrolysates. A maximum yield percentage of 32.1% was obtained for the HCl and HNO_3 acid hydrolysates. Further, the authors also evaluated the production parameters like pH, sugar concentration on cellulose yield. They also reported the physical and chemical characteristics of the developed bacterial cellulose through various analytical evaluations [100].

2.6.1.10 Rice Bark

Rice bark, one of the by-products and wastes created from the rice manufacturing process, is also used as a raw material for bacterial cellulose production. In this study, the researchers used *A. xylinum* as a bacterial strain for cellulose production. Rice bark is hydrolysed with the help of acid and enzyme and then used for the cellulose production. On 10 days of fermentation, higher bacterial cellulose production without the addition of protein and other supplements like yeast is noticed. A yield of 2.42 g/L cellulose is obtained in static production, and 1.57 g/L is noted for the agitated culture. Out of the selected methods for the cellulose production with rice bark, the static method found to be more effective. The researcher further evaluated the physical properties of the cellulose and crystalline nature of the product in detail [101].

2.6.1.11 Corn Steep Liquor

Corn steep liquor is one of the widely discussed materials for bacterial cellulose production than other alternative materials. The corn steep liquor on the cellulose production was analysed, and the role of its constituents was discussed by Matsouka et al. They evaluated the cellulose production capacity of the corn steep liquor along with several other components. The findings of the research revealed that the presence of laccase in the corn steep liquor is one of the main reasons for the higher cellulose production. The researcher supported the findings by mentioning the higher cell growth in the medium than glucose. By analysing different amino acids present in the corn steep liquor along with the laccase, they had developed a synthetic medium for *A. xylinum* sub sp. *Sucrofermentans,* which can yield 90% cellulose production in a jar fermentor [102]. Other researchers used *Gluconacetobacter hansenii* against the corn steep liquor for the production of cellulose. The researcher developed an alternative media by including the corn steep liquor in the HS medium component by replacing few items. They developed different compositions to evaluate the cellulose production capacity of the *G. hansenii.* Out of the selected materials like sugarcane molasses and acetylated glucose, yeast extract, peptone, and corn steep liquor, only corn steep liquor is identified as the potential medium [103].

Similar study used the corn steep liquor with the glycerol obtained from biodiesel production and grape bagasse obtained from the wine production process. They evaluated the cellulose production with *G. xylinus.* The findings revealed that out of combinations tried, namely grape bagasse, grape bagasse + diammonium phosphate, grape bagasse + Corn steep liquor, glycerol biodiesel + Corn steep liquor with 14 days fermentation, grape bagasse + Corn steep liquor, glycerol biodiesel + Corn steep liquor combinations yielded a maximum of 8 and 10 g/L bacterial cellulose. In further analysis, the thermal and crystalline nature of the developed product are detailed by the researchers [104]. A recent review by Zohaib Hussain et al. reported various applications of bacterial cellulose along with different potential raw materials that can be used as an alternative feedstock material for the bacterial cellulose. In that systematic review, the researcher classified the several industrial wastes like agro waste, brewery/beverage industry waste, sugar industry, pulp and biorefinery industry wastes and textile industry wastes as a potent source of bacterial cellulose. To understand and explore further in the alternative carbon sources, this research work will aid good basics for the readers [105]. Table 2.3 presents the different carbon and nitrogen sources reported by previous researchers.

2.7 Role Nitrogen Source on Bacterial Cellulose Production

Nitrogen source is one of the important nutrients for bacterial cellulose production, as it enhances the cell construction and growth of the microorganism. The most proven and most discussed medium for the bacterial cellulose production is HS medium which addresses the use of 0.5% yeast extract and 0.5% peptone by default. Most

Table 2.3 Different sources and media components reported by researchers

S. No	New medium components	New carbon and Nitrogen source used	pH and temperature	References
1	Black tea-5 g, 500 ml water	Yeast extract-5 g Peptone-0.5 g Sucrose-100 g	5, 30 °C	[88]
2	4 g green tea	100 g of sucrose, 1L water, 100 mL of 6% yeast extract	20 ±2 °C	[89]
3	1 L of water, 4 g of green tea	100 ml of 6 per cent yeast extract, 100 g of sucrose	20–24	[90]
4	Grape (Bangalore blue variety), 5% cane sugar	Glucose 20 g/L, peptone 5 g/L, yeast extract 5 g/L, citric acid 1.15 g/L, Di sodium hydrogen phosphate 2.7 g/L	4.5, ambient temperature	[91]
5	RWW (Rice Waste Water)-200 ml, Sucrose-20 g	Urea-1.0 g, glycerol-1.0 g and acetic acid	3–4, room temperature	[92]
6	Coconut milk, sucrose	Sucrose 8%, peptone 0.2%, yeast extract 0.05%, KH_2PO_4 0.1%, $MgSO_4 \cdot 7H_2O$ 0.1%, $(NH_4)_2SO_4$ 0.2%, citric acid 0.1%, 95% ethanol 1.5%, double-distilled water 100% and Sucrose 4%, peptone 0.2%, yeast extract 0.05%, $CaCl_2$ 0.0899%, KH_2PO_4 0.3%, $MgSO_4.7H_2O$ 0.1%, $(NH_4)_2SO_4$ 0.4%, coconut milk 100%, pH 4.5, D-lactic acid 1%, L-lactic acid 0.1%, 95% ethanol 1.5% and sodium acetate 0.8%	4.5, 30 °C	[93]
7	Crude distillery effluent	1.0 N HCl, Sucrose, Corn steep liquor	4.3, 30 °C	[94]
8	Crude glycerol-derived compounds	YDP (yeast extract peptone dextrose) medium—Yeast-10 g/L, Peptone-20 g/L, Agar-15 g/L and dextrose-20 g/L and HS medium (Glucose 20 g/L, peptone 5 g/L, yeast extract 5 g/L, sodium phosphate anhydrous 2.7 g/L and citric acid monohydrate 1.15 g/L	5.0, 30 °C	[96]

(continued)

Table 2.3 (continued)

S. No	New medium components	New carbon and Nitrogen source used	pH and temperature	References
9	Coconut and pineapple juices	1% yeast extract and 14 ml of 95% ethanol in 500 ml	4.75, 30 °C	[97]
10	Nitrogen source-2.0% Peptone, 0.5% yeast extract and 0.12% citric acid	Carbon source-mixed fruit juice which includes orange, apple, pineapple, Japanese pear and grape	6.0, 30 °C	[98]
11	Rotten fruits composed of plums, green grapes, pineapples, and apples and milk whey	Glucose 20 g/L, peptone 5 g/L, yeast extract 5 g/L, sodium phosphate anhydrous 2.7 g/L and citric acid monohydrate 1.15 g/L	6.0, 30 °C	[99]
12	Potato peel waste (PPW)	Glucose-2.0 w/v; peptone-0.5 w/v; yeast extract-0.5 w/v; disodium phosphate- 0.27 w/v; citric acid-0.115 w/v	6.0, 30 °C	[100]
13	Rice bark	Glucose-4 g L^{-1}, peptone 5 g L^{-1}, yeast extract 5 g L^{-1}, Na_2HPO_4 2.7 g L^{-1} and citric acid 1.15 g L^{-1}	5.25, 28 °C	[101]
14	Basal medium	Fructose-4% w/v, KH_2PO_4-0.1% w/v, $MgSO_4 \cdot 7H_2O$-0.025% w/v, Salt solution-1% v/v, Vitamin solution-1% v/v	5.0, 28 °C	[102]
15	Corn steep liquor	Glucose- 2.0 w/v; peptone- 0.5 w/v; yeast extract-0.5w/v; disodium phosphate- 0.27 w/v; citric acid- 0.15 w/v	5.0, 30 °C	[103]
16	Glycerol from biodiesel and grape bagasse (0.5% w/v glycerol) or cane molasses (50.75% w/w fermentable sugars) as carbon source	Corn steep liquor (CSL) or 0.7% w/v diammonium phosphate (DAP) as nitrogen source	5.0, 28 ±1 °C	[104]

of the research works performed on the cellulose production are based on the HS medium. There are different types of the nitrogen sources reported in the previous research by altering the HS medium against the bacterial cellulose production. Su Min Yim et al. analysed the use of various tea substrate as a nitrogen source for the bacterial cellulose production and analysed its properties. The researcher used four different tea, namely black tea, green tea, rooibos tea and corn silk tea, as a nitrogen source. The researchers evaluated the effect of various nitrogen sources on the bacterial cellulose yield, crystallinity and surface characteristics. The findings revealed that green tea is the higher cellulose producing substance than other selected nitrogen sources. The reason behind this is the higher antioxidant activity and lower cell-damaging nature of the green tea. Further, the fibre morphology, surface colour and roughness were evaluated by the researchers. They had reported that green tea produced defined cellulose microfibrils and corn silk tea did not produce any fibrils. The other two substances developed cellulose microfibrils but not up to the level of green tea [106].

Embuscado et al. analysed the effect of various organic and inorganic nitrogen sources on bacterial cellulose production. The researcher developed 24 different combinations of the organic and inorganic nitrogen sources and reported its influence on the bacterial cellulose yield. In general, the organic nitrogen sources like peptone, tryptone and yeast extract produced higher cellulose yield than the inorganic sources like ammonium phosphate, ammonium sulphate and potassium nitrate. They noted a better yield of 5.26 g/L cellulose for the combination of two organic compounds peptone and yeast extract. Further, they reported that the combination of organic and inorganic substances improved the cellulose yield than the inorganic alone. However, with the combination of two inorganic substances, they found very little or no improvement in cellulose production [1]. Ramana et al. evaluated the role of carbon and nitrogen source on the cellulose production using *A. xylinum.* In their analysis, the researcher used peptone, soybean meal, glycine, casein hydrolysate and glutamic acid as nitrogen sources in their study. Based on the study, they found that in the sucrose medium, the cellulose yield was noted higher with the peptone or casein hydrolysate than the soybean meal or ammonium sulphate. This finding supported that the nitrogen source also helps cell enhancement if suitably selected [107].

Natthawut Yodsuwan et al. analysed the effect of various nitrogen sources on the bacterial cellulose production. The researcher used yeast extract, peptone, $(NH_4)_2SO_4$, polypeptone and casein hydrolyte in the sucrose medium to produce bacterial cellulose. Researchers also evaluated the different combinations of nitrogen sources and their effect on cellulose yield. They found that the use of a single nitrogen source reduced the cellulose yield than the combined nitrogen sources. The yeast extract + casein hydrolysate, yeast extract + yeast extract medium and yeast extract + ammonium sulphate medium are the combined mediums, which showed a significant improvement in the cellulose yield [108]. Other researchers tried the corn steep liquor as an alternative medium for cellulose production. Their findings reported that the corn steep liquor contained various amino acids in its structure. However, the researcher did not achieve the same amount of cellulose yield as of corn steep liquor when they use the identified amino acids together and separately. The researcher also reported that the methionine, a type of amino acid, present in the corn steep liquor had an important role in cell growth and cellulose production. In comparison with a media without amino acids, their findings reported a higher yield of 90% with the amino acid [109]. On the bacterial cellulose production with fruit juices, the researcher tried different combinations like fruit juice with a nitrogen source and without nitrogen sources. The results of the research mentioned that the addition of nitrogen source increased the cellulose production 8, 6.5, 4.6 and 3.3 times, respectively, for pear, pineapple, grape and orange juice than the juice medium without the nitrogen source. It was reported that the presence of nitrogen content might have improved the protein and amino acid content and so the cell growth and bacterial cellulose production [98].

Hong-Joo Son et al. analysed the effect of various nitrogen sources on cellulose production using *Acetobacter* sp. The researcher analysed the effect of various nitrogen sources like beef extract, corn steep liquor, malt extract, polypeptone, proteose peptone, tryptone, yeast extract, $(NH_4)_2SO_4$, NH_4Cl, KNO_3 and without the addition of any nitrogen source on the cellulose production. The findings revealed that the media with corn steep liquor (CSL), polypeptone, or yeast extract yielded a higher quantity of cellulose concerning the litre of media. Out of all media used, yeast extract is identified as the most suitable for the selected bacterial strain. The rich nitrogen content in the yeast extract is one of the main reasons for its maximum cellulose yield (2.87 g/L). Further, the researchers reported that 0.1% w/v of yeast extract did yield maximum cellulose, and further, increment in the concentration did not have any significant improvement in the production. At the same time, the researcher also mentioned that the addition of 0.7% w/v of polypeptone increased the bacterial cellulose production 1.5 times higher. Out of the research, the findings reported that the corn steep liquor and yeast extracts are the maximum cellulose production nitrogen source for *Acetobacter* sp. [69].

2.8 Role of Media pH on Bacterial Cellulose Production

pH is one of the important parameters which has a direct impact on bacterial cellulose production. Research findings identified that bacterial cellulose production was affected by lower pH values. Masaoka et al. reported that during the fermentation process, the conversion of glucose into gluconic acid leads to lower pH in the culture medium. They reported that the pH value for the cellulose production from *A. xylinum* was in the range of 4–6 [109]. Other researchers also mentioned that the bacterial cellulose growth increases the acetic acid in the fermentation medium due to the conversion of carbohydrates to acetic acid by *A. xylinum*. The synthesis metabolism is a kind of respiratory process, which converts the ethanol into acetic acid as a result of oxidisation process and alters the glucose into gluconic acid [97]. Hence, it is important to have solid control over the fermentation medium during the bacterial cellulose production. The general pH range for the *A. xylinum* is 6–7 as reported by Hestrin and Schramm [12].

Klemm et al. reported that the main reason for the pH change in the medium is due to the development of gluconic acid and 5-ket-gluconic acid during the initial cultivation. Further to add, the conversion of the monosaccharides as the (keto) gluconic acids was performed by the *Acetobacter* dehydrogenase. This process generally provides a negative impact on cellulose productivity and cell viability for the subsequent synthesis [110]. The pH shift in the cellulose production from coconut and pineapple juices was analysed by other researchers. They have reported that the control of the pH in the cellulose producing medium by using pH buffer system is difficult. Out of the selected juices, the coconut juice medium produced a higher amount of pH change in the medium. They measured the acetic acid contents and mentioned that the coconut juice produced 24.96 times more amount of acid than the initial stage with *A. xylinum.* In the case of pineapple juice, the increment is noted less for all the tested strains. The reduction in the pH to the acidic region not only has a serious effect on cellulose production but also the growth [97]. Kazim reported the effect of pH on the cellulose fermentation with *Pseudomonas* spp. The researcher reported that the maximum production of cellulose obtained at pH value of 5 with a yield of 6.9–4.3 g/L. For the *Pseudomonas* spp., the optimum pH value as reported by the researcher is in the range of 5.25–5.75 and this is the range in which the gluconic acid production is noted very less than other ranges [38].

Though the pH value of 5.4–6.3 is noted as the opt one for the growth of the *Acetobacter* species, the pH for cellulose production is different. The optimum pH value for the higher cellulose production is reported as 4.5–5.5. There is no cellulose growth noted at the lower pH and no production noted at the pH less than 3.5 [1]. Dirisu et al. measured the role of pH on the cellulose production from *G. xylinus* using *Moringa oleifera* tea–yeast extract broth. The researcher analysed the yield with various pH levels and noted a maximum cellulose yield of 0.32 g/L at a pH range of 4–6. At pH 6, a very minimum yield of 0.01 g/L cellulose noted. No cellulose production of cellulose is noted at the pH range of 7–8. The researcher reported the optimum pH of 4–6. As reported by the previous researchers, the pH change was supported by the generation of acid due to the metabolic reaction. The researcher also mentioned that the mediums with higher pH (7–8) at the initial stage showed a reduction of pH after the fermentation of 15 days and the lower pH of range 4 is found to be increasing towards the neutral pH [111].

In the process of producing bacterial cellulose from acidic by-products of the alcohol and dairy industries, researchers used *G. sucrofermentans.* In this process, the effect of pH on the cellulose production evaluated for the specified strain. The results revealed that after 2 days of growth, the pH of the medium changed from 3.9 to 6.5 for thin stillage medium and from 4.9 to 8 in the case of whey medium. The researcher noted a higher cellulose yield of 6.19 g/L at a pH of 3.95 for thin stillage. The acidic pH of the developed medium is noted as one of the advantages of the cost-effective production of bacterial cellulose [112]. On the production of cellulose from the agitated culture, using *Acetobacter* sp. the role of medium pH noted as important. Out of the tested pH range (3–9), the researcher identified that the pH range of 4.5–7.5 is suitable for cellulose production. In particular, with *Acetobacter* sp. a pH of 6.5 yielded a higher cellulose production than the other conditions. The researcher reported that the general cellulose production will occur for *Acetobacter* sp. in the range of 4–7 pH [69].

Jung Wook Hwang analysed the effect of pH on the bacterial cellulose production from *A. xylinum* using a jar fermentor. In the production process, the glucose oxidation activity of the cells is largely affected by the pH of the medium. However, they reported a constant oxidation rate at a single pH. A maximum oxidation activity (value of 1) noted at 4 pH. Similarly, the cell synthesis, glucose oxidising activities at pH of 3.5 and 5 are noted as 0.57 and 0.3. The findings revealed that cellulose synthesis is favoured at a pH of 5, though the conversion of glucose to gluconic acid happens at pH 4 due to the improved cell synthesis. By batch feed method, a cellulose production of 20 g/L obtained by proper control pH in 42 h [113]. In a study, the *Komagataeibacter saccharivorans* strain BC1 isolated from rotten green grapes were used to produce bacterial cellulose. On optimising the medium content for the higher bacterial cellulose yield, the researchers analysed the favourable pH of the cellulose production from a pH of 2–8. Out of all the experiments, the higher cellulose production of 1.82 ± 0.65 g/100 mL was noted at a pH value of 5 [114]. In a similar study, *Acetobacter* sp. was isolated from the waste washing waters from the temple. The researcher used the isolated strains to produce bacterial cellulose using the wastewater as a medium. In this research, the researcher evaluated the role of pH on cellulose production. In contrary to previous researchers, he reported a higher pH as the optimum for maximum cellulose yield. The study performed in the pH range of 4–8 and a maximum cellulose production of 0.81 g/L were obtained at 7 pH [115].

2.9 Role of Oxygen Supply in Bacterial Cellulose Production

The cellulose floatation in the static cultivation method is noted as a function of oxygen pressure. Cellulose production in the upper part of the static culture is purely due to the quality cell growth in the top side of the vessel. This was proved by the firm structure of the cellulose in the top side and soft, transparent cellulose structure under the bottom side of the floatation matrix. Compared to the agitated culture method, static culture method usually preferred in the aspects of bulk production. However, the limited oxygen supply is one of the main drawbacks of the static culture method (submerged cultivation method). The insufficient oxygen supply at the bottom of the formed cellulose pellicle reduces cell growth and creates reduced cellulose production. While the production of bacterial cellulose increases, the oxygen content in the medium gets depleted completely in due course of time. This limitation is reported by several researchers in the bacterial cellulose production while compared to the agitated culture method [12]. The various production methods with different bioreactors (like a stirred tank, rotating disc bioreactor, airlift reactors, etc.) were introduced to increase the oxygen supply for active cell growth. In the case of the static and agitated culture medium, the surface area of the medium is also one of the directly influencing parameters which are linked to the dissolved oxygen availability.

The effect of dissolved oxygen on the cellulose production was analysed by Jung Wook Hwang et al. [113]. In their research, they used a stirred tank reactor for cellulose production and analysed the effect of dissolved oxygen content on cellulose production. In the tested range of 2–15% of dissolved oxygen, 10% dissolved oxygen yielded maximum cellulose of 15.2 g/L. The oxygen concentration was maintained (10%) for a while of 40 h, and a higher production was noted in the batch; however, from 40 to 50 h time, there is a limited cellulose growth noted [113]. The researcher reported that the bacterial cellulose production rate was mainly based on the oxygen transfer coefficient. The oxygen transfer rate and cellulose production are generally higher in the agitated culture than the static culture. The research performed on airlift bioreactor produced a better result by providing increased oxygen to the fermentation medium. In comparison with the normal air supply, oxygen supply increased the cellulose production. For a 67 h fermentation with normal air, cellulose production of 3.8 g/L was obtained and the oxygen-fortified air supply increased the cellulose production to 8 g/L, which is two times higher than the previous production. By these findings, it is confirmed that the dissolved oxygen is one of the important factors in cellulose production [116]. Li et al. evaluated the role of critical oxygen tension on bacterial cellulose production with *A. xylinum* KJ1. They identified that oxygen is one of the essential requirements for aerobic liquid fermentation with *A. xylinum*. In the research with 50L modified bubble reactor, the researcher used different rates of oxygen feed from 5–20 L/min were added at various time intervals up to 72 h. In their analysis, the researcher maintained 2.13 ppm dissolved oxygen concentration for the initial time, and later, it was maintained at 1.12 ppm. The findings revealed that a dissolved oxygen concentration of 3.10 ppm and above should be maintained for better cellulose production. Due to the liquid and film of cellulose developed on the surface, the oxygen transfer rate is majorly interrupted from reaching the cells. Hence, they conclude that the supplement of oxygen is necessary to increase cellulose production [117].

The oxygen tension in the medium was evaluated based on cellulose production and the performance of the cellulose membrane. The findings revealed a negative correlation between cellulose production and oxygen content. At lower oxygen content, they produced cellulose membranes with less amount of cellulose than the cellulose produced at higher oxygen supply. In addition to that, they also reported a direct influence of oxygen supply on the strength of the cellulose membrane developed. An increase of 10% oxygen tension increased the cellulose generation 25% in the medium but at the same time, it does not influence the cell growth [118]. On the production of cellulose from coconut water using *Acetobacter* sp. in agitated culture, a rotating speed of 100 rpm yielded a maximum production of 5.67 g/L. The researcher identified that the major reason for reduced cellulose production at lower rpm should have resulted due to lower dissolved oxygen content and so the lower cell growth. The researcher measured the dissolved oxygen content in the 50, 100 and 150 rpm culture and noted that 150 rpm culture had a higher amount of dissolved oxygen in it. Two contrary situations noted are: (i) at 150 rpm higher dissolved oxygen noted and that produced a higher gluconic acid generation and so yields a reduced cellulose production and (ii) at 50 rpm, there is less dissolved oxygen and that too insufficient for cellulose production. The researchers concluded that the excess amount of dissolved oxygen oxidises the glucose into gluconic acid as a substitute for cellulose production. Hence, the overfeed of oxygen also reduces bacterial cellulose production [119]. Table 2.4 consolidates the various media components and its key role in cellulose production.

2.10 Role of Other Nutrients in Bacterial Cellulose Production

In order to increase cell growth and cellulose production, researchers used several additional components in the production media. Materials like glycerol, industrial waste and polysaccharides are introduced into cellulose production media to increase cell growth. Particularly, several alcohols were tested against the cellulose production capabilities. Out of all, ethanol is noted as one of the highly discussed alcohols with respect to bacterial cellulose production. Takaaki Naritomi et al. used *A. xylinum* for cellulose production with the carbon source fructose. The researcher used ethanol as an additional source in the culture media to increase the production of bacterial cellulose. They have used various concentrations of ethanol from 4 to 10 g/L. The researcher obtained a maximum yield of 0.95 g/L/h for 10 g/L of ethanol feed in the media. A further increase in the ethanol content to 15 g/L created a sudden reduction in cellulose production and also in cell growth. The main reason reported by the author is the use of ethanol alters the biosynthesis pathway by increasing the flow of glucose 6-phosphate, which is one of the precursors for bacterial cellulose production [120].

Lu et al. analysed the use of methanol, ethylene glycol, *n*-propanol, glycerol, *n*-butanol and mannitol at different concentrations from 0.5 to 5.0% v/v. In this research, they reported that out of all the additives tested 1–3% of the methanol, 1–5% of the mannitol and glycerol, 0.5–1% of the ethylene glycerol, *n*-propanol and *n*-butanol are found to increase the bacterial cellulose production. However, a higher concentration of the same alcohols created an inhibitory effect on cellulose production. Along with HS medium, when the production yield was estimated for 6 days, the highest yield of 132.6 mg/100 ml was obtained for n-butanol added medium, which was 56% higher than the control medium tested. The second most promising additive is mannitol, which produced 125.2 mg/100 ml of cellulose, with 47.3% higher production than the control medium. Similarly, ethylene glycol with 21.8%, *n*-propynol with 24.1%, glycerol with 13.4% increments was noted as the control medium [121]. Detailed analysis of the role of ethanol on bacterial cellulose production was performed by Shunji Yunoki et al. In their finding, they reported that the bacterial cellulose production is initially half of the control medium for the first 24 h incubation. In the subsequent fermentation at 48 h both the batch showed a

Table 2.4 Various media components and its key role in cellulose production

S. No	Media component/condition	Key role in cellulose production
1.	Carbon source	(i) Glucose, sucrose and fructose are the common carbon sources used for the bacterial cellulose production (ii) It is one of the essential nutrients for cell growth (iii) The carbon sources identified by the standard medium are costly and so the alternative cheaper sources were identified by considering industrial-scale production (iv) Corn steep liquor, rice bark, different types of tea, industrial wastes, fruit juices, coconut milk, industrial by-products like glycerol are identified and successfully verified for cellulose production
2.	Nitrogen source	(i) Provides nutrition for cell growth and metabolism (ii) Yeast extract and Peptone are the most commonly used nitrogen sources (iii) Hence, it is essential to have a nitrogen source in the media to produce bacterial cellulose (iv) Various alternative nitrogen sources like tea, corn steep liquor, fruit juices were demonstrated for successful cellulose production
3.	pH	(i) The pH of cell growth and bacterial cellulose production is completely different (ii) Common pH for cellulose production is 4–7 (iii) No bacterial cellulose production noted at pH less than 3.5 (iv) The most specific pH for *Acetobacter xylinum* sp. is noted as 5
4.	Oxygen supply	(i) Increases cell metabolic activity and growth and control the cellulose production rate (ii) Insufficient or excess amount of dissolved oxygen leads to a reduction in cell growth and cellulose production (iii) Use of oxygen yields better cellulose production than normal air
5.	Other nutrients	(i) Various organic acids, alcohols, polysaccharide and industrial wastes with high mineral, vitamin and other nutrients used to reduce the cost of the media and enhance the cellulose production (ii) Ethanol found to be most suitable of all as it ferments and generates acetic acid (iii) This favours the cellulose growth by reducing the inhibiting action of gluconic acid formation
6.	Temperature	(i) The optimum temperature helps cell growth and bacterial cellulose production by metabolism (ii) Out of the different temperature ranges from 20 to 45 °C tested, 28–30 °C is noted optimum temperature for cellulose production by various strains, irrespective of the carbon sources used/media types (iii) A temperature lower and higher than that either reduces or produces no cellulose growth

(continued)

Table 2.4 (continued)

S. No	Media component/condition	Key role in cellulose production
7.	Fermentation time	(i) Fermentation time majorly depends upon the final requirement of the bacterial cellulose (ii) Long fermentation time with a static culture method will produce a continuous sheet of bacterial cellulose. The time duration largely varies from 12 to 21 days (iii) Shorter time agitated culture yields cellulose fibrils in the ball shape. The fermentation time varies from 2 to 5 days

similar quantity of bacterial cellulose. Around twofold cellulose productions were noted with ethanol medium at 168 h of fermentation than the control medium. The researcher correlated the production with glucose consumption. On the observation, it is noted that only half of fed glucose is consumed in the ethanol medium than the control where a higher consumption is noted. The detailed analysis reported that the increased production of bacterial cellulose is mainly due to the lower utilisation of glucose. The addition of ethanol reduced the utilisation of the glucose by the *A. xylinum* and increased the bacterial cellulose production. This is one of the main reasons for glucose availability for long-time cellulose production than the control medium. Further during the synthesis, the ethanol is converted into acetic acid and this improves the cellulose production due to the bacterial consumption instead of the gluconic acid production. The production of acetic acid, instead of gluconic acid provided a favourable environment for the *A. xylinum* to produce cellulose. In both the control and ethanol medium, similar cell growth and pH change were noted without any significant changes and confirmed these factors does not influence cellulose production [122].

Agustin et al. used acetic acid and ethanol to analyse the influence of the additives on the cellulose yield with *A. xylinum* and evaluated its physical properties to find better cellulose. On the use of different concentrations of the additives from 0.5 to 2% v/v, researchers found that at 1.5% concentration of ethanol produced cellulose with Young's modulus of 107.089 N and acetic acid produced Young's modulus of 92.183 N [123]. By using Pullulan polysaccharide fermentation wastewater without any further treatment as media component, researchers developed bacterial cellulose with *Gluconacetobacter* sp. The study compared the cellulose production with HS medium and HS medium with wastewater. Upon fermentation, for 10 days the wastewater polysaccharide added medium produced a slightly higher amount of cellulose (1.177 g/L) than the HS medium (1.157 g/L). The analysis of the culture media revealed that the wastewater media showed a great reduction of 18.9% in its chemical oxygen demand after 10 days of fermentation. The oxygen demand reduced from 39,040 to 31,652.5 mg/L. This represents the wastewater can be used as a nutrient for cellulose production. However, the research also reported a significant change in the mechanical properties of the cellulose structure produced in terms of crystallinity, mechanical strength and structure compared to the HS medium [124].

Other researchers used purified acetan or agar in the cellulose production medium with fructose and corn steep liquor to increase the cellulose production. After 120 h fermentation, a cellulose yield of 4.44 g/L is noted in the case of 1.5 g/L acetan added medium and a yield of 4.48 g/L is noted with 1.5 g/L agar added medium. The corn steep liquor and fructose medium were used as a control medium, which produced a yield of 2.88 g/L cellulose. To evaluate the role of acetan in cellulose production, the researchers used a mutant *A. xylinum,* which cannot produce acetan. The findings revealed that the culture with acetan and agar provided a higher viscosity in the initial stage and this aided an increased cellulose production by reducing coagulation of bacterial cellulose and cells in the culture [125].

The effect of organic acids like oxalic acid, tartaric acid, pyruvic acid, malic acid, lactic acid, acetic acid, citric acid and succinic acid on the bacterial cellulose was evaluated. Upon 0.1% of the addition of these organic acids in the culture medium, the effect on cellulose production was evaluated. The findings revealed that oxalic acid and tartaric acid did not yield any significant amount of cellulose in the medium. Other than that, malic acid, pyruvic acid and citric acid yielded a production of 2.83 g/L, 2.34 g/L, 2.27 g/L with *A. xylinum.* This production amount was a bit higher than the medium which contains ethanol (1.93 g/L). In the case of lactic acid, acetic acid and succinic acid, there is a less amount of cellulose production noted as 1.67 g/L, 1.85 g/L and 1.49 g/L respectively [126]. Zhou et al. evaluated the suitability of sodium alginate on bacterial cellulose production. Their results reported that the addition of sodium alginate up to 0.4% had a positive impact on the cellulose production and the maximum of 6.0 g/L cellulose is obtained from the medium. However, they reported no increment at a higher concentration of sodium alginate than 0.4%. They reported a lag in the first 24 h of cellulose production with a control medium, and it was eliminated in sodium alginate added medium. This was mainly due to the hindrance action of the sodium alginate against the formation of large clumps of bacterial cellulose [127].

2.11 Role of Temperature on Bacterial Cellulose Production

Temperature is one of the other important parameters which has a direct influence on both the cell metabolism and cellulose production of the bacterial strain. Hong-Joo Son et al. evaluated the role of temperature on the cellulose production using *Acetobacter* sp. They analysed in the range of 20-40 °C and noted that around 25–30 °C there is a similar amount of cellulose production noted. At temperature above 35 °C, the production of cellulose got reduced significantly [69]. Jonas et al. reported that an optimum temperature of the bacterial cellulose production is in the range of 24–30 °C. At 24 °C, the researcher obtained a 50% better cellulose yield than any other temperature. [22]. In a similar study, Santosh Udhavrao Pokalwar obtained a better yield of cellulose from *Gluconacetobacter* intermedius at 30 °C (4.0 g/L). The researcher identified a lower production of 3.2 g/L, 3.0 g/L and 2.0 g/L respectively, for the temperatures 35 °C, 25 °C and 20 °C after 144 h of incubation. They had reported a reduction in cellulose production at both ends, on higher and lower sides of the temperature 30 °C [128]. On the production of cellulose from *A. xylinum* using the pineapple waste, the researcher analysed the role of temperature and found that the cellulose production increased with the increase in temperature from 28 to 30 °C. And there is a sudden reduction in the cellulose yield at 31 °C. The researcher optimised the best yielding cellulose using response surface methodology and reported a temperature of 30.59 °C as the best for selected strain and the medium [129]. Concerning *Gluconacetobacter* strains, other researchers evaluated the viability in the range of 20–45 °C. Their results reported that maximum production of 2.31 g/L was obtained at the temperature range of 28–30 °C. The least production is noted at 37 °C and no production is observed in the range of 40–45 °C [130].

2.12 Role of Fermentation Time on Bacterial Cellulose Production

The static culture method is normally preferred for industrial-scale bacterial cellulose production due to the effectiveness of this method in cellulose production as continuous sheet. However, one of the basic drawbacks of the static culture method is its long-term fermentation. The long fermentation time varies from 18 to 21 days, which is usually used to produce cellulose in the form of three-dimensional matrixes. However, the cellulose can be produced with proper media and bacterial strain within 2–3 days. Research works performed to evaluate the cellulose production possibilities and focused on the measurement of cellulose productivity or yield percentage were used a shorter period of 2–3 days. However, research works on specific applications required cellulose matrix sheet and biosynthesis-based research should be carried out till the end of fermentation. These research works were generally performed for a longer duration. Cellulose formation can be noted in the first few days of incubation but the matured matrixes can be obtained only in the case of long-term fermentation. Goh et al. evaluated the role of fermentation time on the fermentation of kombucha and reported that upon 18 days fermentation, the tea broth produced 199.3 g/L bacterial cellulose with 2 cm thickness. The production of cellulose is nil after the 18th day of fermentation [131]. The reason behind the halt of the bacterial cellulose is the exhaustion of the sucrose content in the medium. In the initial incubation period, the cell in the medium used the dissolved oxygen for its growth and produces cellulose and after a due course of time, upon the depletion of dissolved oxygen, the growth of the cellulose in the fermented medium depends on the surface oxygen. In this period, the bacterial cells directly in contact with the air and produce cellulose effectively but the cells in the bottom side of the fermentation tank do not have the option to get enough oxygen to produce cellulose. They also reported that these bacterial strains are not dead and they are asleep and can be used for future cultures [132].

Several researchers performed research to reduce seed culture preparation time, cell cultivation and cellulose production time through different modified reactors and methods. Similarly, the additions of extra nutrients like alcohol and acids also evaluated by different studies to reduce the fermentation time. A study by Nur Arfa Yanti et al. analysed the effect of fermentation time on the bacterial cellulose yield with sago liquid waste obtained from the sago starch processing industry. Their cellulose production with *A. xylinum* LKN6 was analysed for a fermentation period of 20 days. The results revealed that a maximum cellulose yield of 13.85 g/L noted in 15 days. Further increment in the fermentation time to 20 days slightly reduced the amount of cellulose production to 13.41 g/L [133]. To optimise the several factors, Sangok Bae and Makoto Shoda used Box—Behnken design and a 10 L Jar fermentor reaction chamber. They have used fructose, corn steep liquor medium for the cellulose production and agar as an additional nutrient. The findings revealed that a maximum of 14.3 g/L bacterial cellulose can be produced with the following parameters namely, fructose—4.99%; corn steep liquor—2.85%; dissolved oxygen—28.33%; and 0.38% agar. These optimised parameters increased the cellulose production in a shorter fermentation time of three days [73]. Other study evaluated the potential of *Gluconacetobacter persimmonis* to produce bacterial cellulose in short fermentation time. A maximum yield of 6.71 g/L cellulose was obtained using fermenter in incubation time of five days, due to the increased oxygen supply. But in the ideal case with static culture, a cellulose yield of 5.89 g/L was obtained in 12–14 days [134]. Sasithorn Kongruang evaluated the cellulose production using different strains, namely *A. xylinum* TISTR 998, *A. xylinum* TISTR 975 and *A. xylinum* TISTR 893, using coconut and pineapple juices. In their research, a maximum yield of bacterial cellulose noted against 14 days fermentation [97]. Table 2.5 details the different long fermentation time followed for various bacterial strains by other researchers for the production bacterial cellulose 3D matrix.

2.13 Summary

This chapter provided an outline of the various mechanisms involved in bacterial cellulose production from the different types of bacteria to the raw material used. *A. xylinum* and its family (*Acetobacter* Sp.) were highly discussed in terms of its cellulose production mechanism, cultivation media and method due to its higher cellulose production capacity. The mechanism and reason for bacterial cellulose production by different species are also outlined clearly. Out of conventional and unconventional production methods adapted, majorly used industrial reactors and methods are detailed with its merits and demerits. The media component is one of the essential aspects of bacterial cellulose production and at the same time one of the huge barriers for commercialisation. Hence, the last portion of the chapter details the influence of various media components on cell growth, bacterial cellulose and its

Table 2.5 Fermentation time of different bacterial strains to produce bacterial cellulose

S. No	Bacterial Strain	Type of medium used	Fermentation time (days)	References
1.	*Acetobacter xylinum*	HS medium	18–20	[109]
2.	*Acetobacter xylinum*	HS medium	14	[135]
3.	*Acetobacter xylinum*	HS medium + Rice bark	10	[101]
4.	Kombucha-based *Acetobacter xylinum*	Tea	12	[106]
5.	*Acetobacter xylinum*	HS-Baikal EM1 medium	14	[136]
6.	*Gluconacetobacter hansenii*	HS + Corn steep liquor medium	10	[103]
7.	*Komagataeibacter xylinus*	date syrup and cheese whey + HS	12	[137]
8.	Kombucha-based *Acetobacter xylinum*	Black Tea	28	[138]
9.	*Acetobacter xylinum*	HS medium	21	[139]
10.	*Kombucha scoby*	Tea medium	21	[140]

importance of bacterial metabolism. Along with that, the chapter also summarised a few of the alternative cost-effective culture media tried by various researchers to enlighten the commercial potential of the bacterial cellulose for future applications.

References

1. Embuscado ME, Marks JS, Bemiller JN (1994) Bacterial cellulose. Optimization of cellulose production by *Acetobacter xylinum* through response surface methodology. Food Hydrocoll 8(5):419–430
2. Brown AJ (1886) On an acetic acid ferment which forms cellulose. J Chem Soc 49:432–439
3. Vergara BS, Idowu PMH, Sumangil JH (1999) Nata de coco—a Filipino delicacy. National academy of Science and Technology, Island publishing house, Inc. Manila, Philippines. http://scinet.dost.gov.ph/union/Downloads/nast_Vergara%20BS%20Nata%20De%20Coco_422.pdf
4. Sharangi AB, Datta S (2015) Value addition of horticultural crops: recent trends and future directions. Springer, Berlin, p 151
5. Lapuz MM, Gallardo EG, Palo MA (1969) The nata organism-cultural requirements, characteristics, and identity. Philippine J Sci 96:91–109
6. Iguchi M, Mitsuhashi S, Ichimura K, Nishi Y, Uryu M, Yamanaka S, Watanabe K (1988) Bacterial cellulose-containing molding material having high dynamic strength, US Patent 4,742,164
7. Yamanaka S, Watanabe K, Kitamura N, Iguchi M, Mitsuhashi S, Nishi Y, Uryu M (1989) The structure and mechanical properties of sheets prepared from bacterial cellulose. J Mater Sci 24:3141–3145

8. Nishi Y, Uryu M, Yamanaka S, Watanabe K, Kitamura K, Iguchi M, Mitsuhashi S (1990) The structure and mechanical properties of sheets prepared from bacterial cellulose. Part 2: improvement of the mechanical properties of sheets and their applicability to diaphragms of electro-acoustic transducers. J Mater Sci 25:2997–3001
9. Sisson WA, Clark GL (1933) X-ray method for quantitative comparison of crystalline orientation in cellulose fibers, Ind Eng Chem (Anal. ed.) 5:296
10. Franz E, Schiebold E (1943) Contributions to the structure of bacterial cellulose. J Makromol Chem 1:4
11. Schramm M, Hestrin S (1954) Synthesis of cellulose by *Acetobacter xylinum*. I. Micromethod for the determination of celluloses. Biochem J 56(1):163–166
12. Schramm M, Hestrin S (1954) Factors affecting production of cellulose at the air/liquid interface of a culture of *Acetobacter xylinum*. J Gen Microbiol 11(1), 123–129. https://doi.org/10.1099/00221287-11-1-123
13. Brown RM Jr, Martin WillisonJ H, Richardson Carol L (1976) Cellulose biosynthesis in *Acetobacter xylinum*: visualization of the site of synthesis and direct measurement of the in vivo process. Proc Nat Acad Sci USA 73(12):4565–4569
14. Brown RM Jr (1985) The microbial synthesis of cellulose. In: BIOEXPO85. Kahners Exposition Group, Boston, Massachusetts, pp 325–335
15. Lin FC, Brown RM Jr, Cooper J, Delmer D (1985) Synthesis of fibrils in vitro by a solubilized cellulose synthase from *Acetobacter xylinum*. Science 230:822–825
16. Brown RM Jr (1989) Cellulose biogenesis and a decade of progress: a personal perspective. In: Schuerch C (ed) Cellulose and wood chemistry and technology. Wiley, New York, pp 639–657
17. Brown RM Jr (1989) Bacterial cellulose. In: Kennedy, Phillips, Williams (ed) Cellulose: structural and functional aspects. Ellis Horwood Ltd., pp 145–151
18. Shibazaki H, Kuga S, Onabe F, Brown RM Jr (1994) Mechanism of chain folding in bacterial cellulose. In: Proceedings of International Symposium on Fiber Science and Technology. The Society of Fiber Science and Technology, Japan, p 190
19. Saxena IM, Brown RM Jr (1989) Study of cellulose biosynthesis in *Acetobacter xylinum*: a genetic approach. In: Schuerch C (ed) Cellulose and wood-chemistry and technology. Wiley, New York, pp 537–557
20. Benziman M, Haigler CH, Brown RM Jr, White AR, Cooper KM (1980) Cellulose biogenesis: polymerization and crystallization are coupled processes in *Acetobacter xylinum*. Proc Nat Acad Sci USA 77:6678–6682
21. Iqbal HMN (2015) Development of bio-composites with novel characteristics through enzymatic grafting. PhD Thesis (Doctoral dissertation), University of Westminster, London, UK
22. Jonas R, Farah LF (1998) Production and application of microbial cellulose. Polym Degrad Stab 59(1–3):101–106
23. Klemm D, Heublein B, Fink HP, Bohn A (2005) Cellulose: fascinating biopolymer and sustainable raw material. Angew Chem Int Ed 44(22):3358–3393
24. Yu X, Atalla RH (1996) Production of cellulose II by *Acetobacter xylinum* in the presence of 2,6-dichlorobenzo nitrile. Int J Biol Macromol 19:145–146
25. Keshk S, Sameshima K (2006) Influence of lignosulfonate on crystal structure and productivity of bacterial cellulose in a static culture. Enzym Microb Technol 40(1):4–8
26. Shoda M, Sugano Y (2005) Recent advances in bacterial cellulose production. Biotechnol Bioprocess Eng 10:1–8
27. Kamide K, Matsuda Y, liJima H, Okajima K (1990) Effect of culture conditions of acetic acid bacteria on cellulose biosynthesis. Br Polym J 22:167–171
28. Aloni Y, Colen R, Benziman M, Delmer D (1983) Solubilization of the UDP-glucose:1,4-beta-D-glucan 4-beta-D-glucosyltransferase (cellulose synthase) from *Acetobacter xylinum*. A comparison of regulatory properties with those of the membrane-bound form of the enzyme. J Biol Chem 258:4419–4423

29. Brown RM Jr, Montezinos D (1976) Cellulose microfibrils: visualization of biosynthetic and orienting complexes in association with the plasma membrane. Proc Nat Acad Sci USA 73:143
30. Gooday GW (1983) Microbial polysaccharides. In: Buschell ME (ed). Elsevier, Amsterdam, p 85
31. Matthysse AG, Holmes KV, Gurlitz RHG (1981) Elaboration of cellulose by *Agrobacterium tumefaciens* during attachment to carrot cells. J Bacteriol 145:583–595
32. Deinema MH, Zevenhuizen LPTM (1971) Formation of cellulose fibrils by gram-negative bacteria and their role in bacterial flocculation. Arch Mikrobiol 78:42–57
33. Ross J, Mayer Raphael, Benziman Moshe (1991) Cellulose biosynthesis and function in bacteria. Microbiol Rev 55(1):35–58
34. Smit G, Kijne JW, Lugtenberg BJJ (1986) Correlation between extracellular fibrils and attachment of *Rhizobium leguminosarum* to pea root hair tips. J Bacteriol 168:821–827
35. Canale-Parola E, Wolfe RS (1964) Synthesis of cellulose by *Sarcina ventriculi*. Biochim Biophys Acta 82:403–405
36. Canale-Parola E, Borasky R, Wolfe RS (1961) Studies on *Sarcina ventriculi*. III. Localization of cellulose. J Bacteriol 81(2):311–318
37. French AD (1985) Physical and theoretical methods for determining the supramolecular structure of cellulose. In: Nevell RP, Zeronian SH (eds) Cellulose chemistry and its applications. Ellis Horwood Ltd., Chichester, England, pp 84–111
38. Kazim AR (2015) Production, optimization, and characterization of cellulose produced from *Pseudomonas* spp. World J Exp Biosci 3:89–93
39. Hsieh JT, Wang MJ, Lai JT et al (2016) A novel static cultivation of bacterial celluloseproduction by intermittentfeeding strategy. J Taiwan Inst Chem E 63:46–51
40. Hestrin S, Schnerm A, Mager J (1947) Synthesis of cellulose by resting cells of *Acetobacter xylinum*. Nature Lond 159:64
41. Son HJ, Kim HG, Kim KK, Kim HS, Kim YG, Lee SJ (2003) Increased production of bacterial cellulose by *Acetobacter* sp. V6 in synthetic media under shaking culture conditions. Bioresour Technol 86(3):215–219
42. Guhados G, Wan W, Hutter JL (2005) Measurement of the Elastic modulus of single bacterial cellulose fibers using atomic force microscopy. Langmuir 21:6642–6646
43. Mohite BV, Kamalja KK, Patil SV (2012) Statistical optimization of culture conditions for enhanced bacterial cellulose production by *Gluconoacetobacter hansenii* NCIM 2529. Cellulose 19:1655–1666
44. Hornung M, Ludwig M, Schmauder HP (2007) Optimizing the production of bacterial cellulose in surface culture: a novel aerosol bioreactor working on a fed batch principle (Part 3). Eng Life Sci 7:35–41
45. Lin Shin-Ping, Hsieh Shu-Chen, Chen Kuan-I, Demirci Ali, Cheng Kuan-Chen (2014) Semi-continuous bacterial cellulose production in a rotating disk bioreactor and its materials properties analysis. Cellulose 21:835–844
46. Chao Y, Ishida T, Sugano Y, Shoda M (2000) Bacterial cellulose production by *Acetobacter xylinum* in a 50-L internal-loop airlift reactor. Biotechnol Bioeng 68:345–352
47. Wu SC, Li MH (2015) Production of bacterial cellulose membranes in a modified airlift bioreactor by *Gluconacetobacter xylinus*. J Biosci Bioeng. http://dx.doi.org/10.1016/j.jbiosc.2015.02.018
48. Reiniati I, Hrymak AN, Margaritis A (2017) Kinetics of cell growth and crystalline nanocellulose production by *Komagataeibacter xylinus*. Biochem Eng J 127:21–31
49. Jung Jae Yong, Khan Taous, Park Joong Kon, Chang Ho Nam (2007) Production of bacterial cellulose by *Gluconacetobacter hansenii* using a novel bioreactor equipped with a spin filter. Korean J Chem Eng 24(2):265–271
50. Babaeipour V, Bagherniya M, Soleimani A (2020) Optimization of bacterial nano-cellulose production in bench-scale rotating biological contact bioreactor by response surface methodology. Iran J Chem Chem Eng (IJCCE). http://www.ijcce.ac.ir/article_38040.html
51. Fijalkowski K, Żywicka A, Drozd R et al (2017) Increased water content in bacterial cellulose synthesized under rotating magnetic fields. Electromagn Biol Med 36(2):192–201

52. Hong F, Wei B, Chen L (2015) Preliminary study on biosynthesis of bacterial nanocellulose tubes in a novel double-silicone-tube bioreactor for potential vascular prosthesis. BioMed Res Int, 560365. http://dx.doi.org/10.1155/2015/560365
53. Lu H, Jiang X (2014) Structure and properties of bacterial cellulose produced using a trickling bed reactor. Appl Biochem Biotechnol 172(8):3844–3861
54. Cheng KC, Catchmark JM, Demirci A (2011) Effects of CMC addition on bacterial cellulose production in a biofilm reactor and its paper sheets analysis. Bio Macromol 12(3):730–736
55. Hofinger M, Bertholdt G, Weuster-Botz D (2011) Microbial production of homogeneously layered cellulose pellicles in a membrane bioreactor. Biotechnol Bioeng 108(9):2237–2240. https://doi.org/10.1002/bit.23162
56. Cheng Kuan-Chen, Catchmark Jeff M, Demirci Ali (2009) Enhanced production of bacterial cellulose by using a biofilm reactor and its material property analysis. J Biol Eng 3:12. https://doi.org/10.1186/1754-1611-3-12
57. Song HJ, Li H, Seo JH, Kim MJ, Kim SJ (2009) Pilot-scale production of bacterial cellulose by a spherical type bubble column bioreactor using saccharified food wastes. Korean J Chem Eng 26(1):141–146. https://doi.org/10.1007/s11814-009-0022-0
58. Kim YJ, Kim JN, Wee YJ, Park DH, Ryu HW (2007) Bacterial cellulose, production by *Gluconacetobacter* sp. PKY5 in a rotary biofilm contactor. Appl Biochem Biotechnol 137–140(1–12):529–537
59. Bae SO, Shoda M (2005) Production of bacterial cellulose by *Acetobacter xylinum* BPR2001 using molasses medium in a jar fermentor. Appl Microbiol Biotechnol 67:45–51
60. Borines Myra G, Rollon Annalyn P, Barraquio Wilfredo L, Demafelis Rex B, Jose Wilfredo I (2005) Cellulose production by *Acetobacter xylinum* using rotary cylindrical bioreactor. Philippine J Agric Biosyst Eng 3(1):3–11
61. Serafica G, Mormino R, Bungay H (2002) Inclusion of solid particles in bacterial cellulose. Appl Microbiol Biotechnol 58(6):756–760
62. Krystynowicz A, Czaja W, Wiktorowska-Jezierska A, Goncalves- M, Iskiewicz M, Turkiewicz M, Bielecki S (2002) Factors affecting the yield and properties of bacterial cellulose. J Ind Microbiol Biotechnol 29(4):189–195. https://doi.org/10.1038/sj.jim.7000303
63. Chao YP, Sugano Y, Kouda T, Yoshinaga F, Shoda M (1997) Production of bacterial cellulose by *Acetobacter xylinum* with an air-lift reactor. Biotechnol Tech 11(11):829–832. https://doi.org/10.1023/a:1018433526709
64. Yoshino T, Asakura T, Toda K (1996) Cellulose production by *Acetobacter pasteurianus* on silicone membrane. J Ferment Bioeng 81(1):32–36. https://doi.org/10.1016/0922-338x(96)83116-3
65. Han Nam Soo, Robyt John F (1998) The mechanism of *Acetobacter xylinum* cellulose biosynthesis: direction of chain elongation and the role of lipid pyrophosphate intermediates in the cell membrane. Carbohyd Res 313:125–133
66. Brown RM Jr (1987) The biosynthesis of cellulose. Food Hydrocolloids 1:345–351
67. Chawla Prashant R, Bajaj Ishwar B, Survase Shrikant A, Singhal Rekha S (2009) Microbial cellulose: fermentative production and applications. Food Technol Biotechnol 47(2):107–124
68. Naritomi T, Kouda T, Yano H, Yoshinaga F (1998) Effect of lactate on bacterial cellulose production from fructose in continuous culture. J Ferment Bioeng 85:89–95
69. Son HJ, Heo MS, Kim YG, Lee SJ (2001) Optimization of fermentation conditions for the production of bacterial cellulose by a newly isolated *Acetobacter* sp. A9 in shaking cultures. Biotechnol Appl Biochem 33:1–5
70. Ishihara M, Matsunaga M, Hayashi N, Tiler V (2001) Utilization of D-xylose as carbon source for production of bacterial cellulose. Enzyme Microb Technol 31:986–991
71. Park JK, Jung JY, Park YH (2003) Cellulose production by *Gluconacetobacter hansenii* in a medium containing ethanol. Biotechnol Lett 25:2055–2059
72. Son HJ, Kim HG, Kim KK, Kim HS, Kim YG, Lee SJ (2003) Increased production of bacterial cellulose by *Acetobacter* sp. V6 in synthetic media under shaking culture conditions. Bioresour Technol 86:215–219

73. Bae S, Shoda M (2005) Statistical optimization of culture conditions for bacterial cellulose production using Box-Behnken design. Biotechnol Bioeng 90:20–28
74. Sun DP, Zhang JD, Zhou LL, Zhu MY, Wu QH, Xu CY (2005) Production of bacterial cellulose with *Acetobacter xylinum* 1.1812 fermentation. J Nanjing Univ Sci Technol 29:601–604
75. Kim SY, Kim JN, Wee YJ, Park DH, Ryu HW (2006) Production of bacterial cellulose by *Gluconacetobacter* sp. RKY5 isolated from persimmon vinegar. Appl Biochem Biotechnol 131:705–715
76. Seto A, Saito Y, Matsushige M, Kobayashi H, Sasaki Y, Tonouchi N, Tsuchida T, Yoshinaga F, Ueda K, Beppu T (2006) Effective cellulose production by a coculture of *Gluconacetobacter xylinus* and *Lactobacillus mali*. Appl Microbiol Biotechnol 73:915–921
77. Premjet S, Premjet D, Ohtani Y (2007) The effect of ingredients of sugar cane molasses on bacterial cellulose production by *Acetobacter xylinum* ATCC 10245. Sen-i Gakkaishi 63:193–199
78. Nguyen VY, Flanagan B, Gidley MJ, Dykes GA (2008) Characterization of cellulose production by a *Gluconacetobacter xylinus s*train from kombucha. Curr Microbiol 57:449–453
79. Tan L, Ren L, Cao Y, Chen X, Tang X (2012) Bacterial cellulose synthesis in Kombucha by *Gluconacetobacter* sp. and *Saccharomyces* sp. Adv Mater Res 554–556:1000–1003
80. Kose R, Sunagawa N, Yoshida M et al (2013) One-step production of nanofibrillated bacterial cellulose (NFBC) from waste glycerol using *Gluconacetobacter intermedius* NEDO-01. Cellulose 20:2971–2979
81. Hassan EA, Abdelhady HM, El-Salam SSA, Abdullah SM (2015) The characterization of bacterial cellulose produced by *Acetobacter xylinum* and *Komgataeibacter saccharovorans* under optimized fermentation conditions. Brit Microbiol Res J 9:1–13
82. Agustin YE, Padmawijaya KS, Rixwari HF, Yuniharto VAS (2017) Glycerol as an additional carbon source for bacterial cellulose synthesis. IOP conference series: earth and environmental science, vol 141. In: 2nd international conference on biomass: toward sustainable biomass utilization for industrial and energy applications, Bogor, Indonesia, 24–25 Jul 2017
83. Volova TG, Prudnikova SV, Sukovatyi AG et al (2018) Production and properties of bacterial cellulose by the strain *Komagataeibacter xylinus* B-12068. Appl Microbiol Biotechnol 102:7417–7428
84. Fontana JD, Franco VC, Desouza SJ, Lyra IN, De SouzaA M (1991) Nature of plant stimulators in the production of Acetobacter xylinum (tea fungus) biofilm used in skin therapy. Appl Biochem Biotechnol 28–9:341–351
85. Reiss J (1994) Influence of different sugars on the metabolism of the tea fungus. Zeitschrift für Lebensmittel-Untersuchung und –Forschung. 198:258–261
86. Jayabalan Rasu, Malbaˇsa Radomir V, Lonˇcar Eva S, Vitas Jasmina S, Sathishkumar Muthuswamy (2014) A review on kombucha tea—microbiology, composition, fermentation, beneficial effects, toxicity, and tea fungus. Compr Rev Food Sci Food Saf 13:538–550
87. Zhang Wen, Wang Xuechuan, Qi Xiangjun, Ren Longfang, Qiang Taotao (2018) Isolation and identification of a bacterial cellulose synthesizing strain from kombucha in different conditions: *Gluconacetobacter xylinus* ZHCJ618. Food Sci Biotechnol 27(3):705–713
88. Zhu Changlai, Li Feng, Zhou Xinyang, Lin Lin, Zhang Tianyi (2014) Kombucha-synthesized bacterial cellulose: preparation, characterization, and biocompatibility evaluation. J Biomed Mater Res Part A 102A:1548–1557
89. Sederavičiūtė F, Bekampienė P, Domskienė J (2019) Effect of pretreatment procedure on properties of Kombucha fermented bacterial cellulose membrane. Polym Testing 78:105941
90. Domskiene J, Sederaviciute F, Simonaityte J (2019) Kombucha bacterial cellulose for sustainable fashion. Int J Clothing Sci Technol 31(5):644–652. https://doi.org/10.1108/IJCST-02-2019-0010
91. Usha Rani M, Udayasankar K, Anu Appaiah KA (2011) Properties of bacterial cellulose produced in grape medium by native isolate *Gluconacetobacter* Sp. J Appl Polym Sci 120:2835–2841
92. Rohaeti E, Endang Widjajanti Laksono FX, Rakhmawati A (2017) Characterization and the activity of bacterial cellulose prepared from rice waste water by addition with glycerol and chitosan, ARPN J Agric Biol Sci 12(8):241–248

93. Liu D-m Sr, Yao K, Li J-h, Huang Y-y, Brennan CS, Chen S-m, Wu H, Zeng X-A Sr, Brennan M, Li L Sr (2019) The effect of ultraviolet modification of *Acetobacter xylinum* (CGMCC No. 7431) and the use of coconut milk on the yield and quality of bacterial cellulose. Int J Food Sci Technol 54(11):3099–3108. https://doi.org/10.1111/ijfs.14225
94. Jahan Firdaus, Kumar Vinod, Saxena RK (2018) Distillery effluent as a potential medium for bacterial cellulose production: a biopolymer of great commercial importance. Biores Technol 250:922–926
95. Tsouko Erminda, Kourmentza Constantina, Ladakis Dimitrios, Kopsahelis Nikolaos, Mandala Ioanna, Papanikolaou Seraphim, Paloukis Fotis, Alves Vitor, Koutinas Apostolis (2015) Bacterial cellulose production from industrial waste and by-product streams. Int J Mol Sci 16:14832–14849
96. Yang Ho Jin, Lee Taek, Kim Jung Rae, Choi Yoon-E, Park Chulhwan (2019) Improved production of bacterial cellulose from waste glycerol through investigation of inhibitory effects of crude glycerol-derived compounds by *Gluconacetobacter xylinus*. J Ind Eng Chem 75:158–163
97. Kongruang Sasithorn (2008) Bacterial cellulose production by *Acetobacter xylinum* strains from agricultural waste products. Appl Biochem Biotechnol 148:245–256
98. Kurosumi Akihiro, Sasaki Chizuru, Yamashita Yuya, Nakamura Yoshitoshi (2009) Utilization of various fruit juices as carbon source for production of bacterial cellulose by *Acetobacter xylinum* NBRC 13693. Carbohyd Polym 76:333–335
99. Jozala AF, Pértile RAN, dos Santos CA, de Carvalho Santos-Ebinuma V, Seckler MM, Gama FM, Pessoa A Jr (2015) Bacterial cellulose production by *Gluconacetobacter xylinus* by employing alternative culture media. Appl Microbiol Biotechnol 99:1181–1190
100. Abdelraof M, Hasanin MS, El -Saied H (2019) Ecofriendly green conversion of potato peel wastes to high productivity bacterial cellulose. Carbohyd Polym 211(1):75–83. https://doi.org/10.1016/j.carbpol.2019.01.095
101. Goelzer FDE, Faria-Tischer PCS, Vitorino JC, Sierakowski M-R, Tischer CA (2009) Production and characterization of nanospheres of bacterial cellulose from *Acetobacter xylinum* from processed rice bark. Mat Sci Eng C 29(2):546–551. https://doi.org/10.1016/j.msec.2008.10.013
102. Matsuoka M, Tsuchida T, Matsushita K, Adachi O, Yoshinaga F (1996) A synthetic medium for bacterial cellulose production by *Acetobacter xylinum* subsp. sucrofermentans. Biosci Biotechnol Biochem 60(4):575–579
103. Costa AFS, Almeida FCG, Vinhas GM, Sarubbo LA (2017) Production of bacterial cellulose by *Gluconacetobacter hansenii* using corn steep liquor as nutrient sources. Front Microbiol 8:2027. https://doi.org/10.3389/fmicb.2017.02027
104. Vazquez A, Foresti ML, Cerrutti P, Galvagno M (2012) Bacterial cellulose from simple and low cost production media by *Gluconacetobacter xylinus*. J Polym Environ 21(2):545–554. https://doi.org/10.1007/s10924-012
105. Hussain Zohaib, Sajjad Wasim, Khan Taous, Wahid Fazli (2019) Production of bacterial cellulose from industrial wastes: a review. Cellulose 26:2895–2911
106. Yim Su Min, Song Ji Eun, Kim Hye Rim (2017) Production and characterization of bacterial cellulose fabrics by nitrogen sources of tea and carbon sources of sugar. Process Biochem 59:26–36
107. Ramana KV, Tomar A, Singh Lokendra (2000) Effect of various carbon and nitrogen sources on cellulose synthesis by *Acetobacter xylinum*. World J Microbiol Biotechnol 16:245–248
108. Yodsuwan N, Owatworakit A, Ngaokla A, Tawichai N, Soykeabkaew N (2012) Effect of carbon and nitrogen sources on bacterial cellulose production for bionanocomposite materials, Conference paper. https://www.researchgate.net/publication/280730054
109. Masaoka Satoshi, Ohe Tatsuhiko, Sakota Naokazu (1993) Production of cellulose from glucose by *Acetobacter xylinum*. J Ferment Bioeng 75(1):18–22
110. Klemm D, Schumann D, Udhardt U, Marsch S (2001) Bacterial synthesized cellulose artificial blood vessels for microsurgery. Prog Polym Sci 26:1561–1603

111. Dirisu CG, Rosenzweig J, Lambert E, Oduah AA (2017) pH effect and pH changes during biocellulose production by *Gluconacetobacter xylinus* in *Moringa oleifera tea*-sugar medium. J Adv Microbiol 7(2):1–7
112. Revin V, Liyaskina E, Nazarkina M, Bogatyreva A, Shchankin M (2018) Cost-effective production of bacterial cellulose using acidic food industry by-products. Braz J Microbiol 49(Suppl 1):151–159
113. Hwang Jung Wook, Yang Young Kook, Hwang Jae Kwan, Pyun Yu Ryang, Kim Yu Sam (1999) Effects of pH and dissolved oxygen on cellulose production by *Acetobacter xylinum* BRC5 in agitated culture. J Biosci Bioeng 88(2):183–188
114. Gopu G, Govindan S (2018) Production of bacterial cellulose from *Komagataeibacter saccharivorans* strain BC1 isolated from rotten green grapes. Prep Biochem Biotechnol 48(9):842–852. https://doi.org/10.1080/10826068.2018.1513032
115. Raghunathan D (2013) Production of microbial cellulose from the new bacterial strain isolated from temple wash waters. Int J Curr Microbiol Appl Sci 2(12):275–290
116. Chao Y (2002) Characteristics of bacterial cellulose production by *Acetobacter xylinum* by an airlift reactor. Ph.D. Thesis. Tokyo Institute of Technology, Tokyo, Japan
117. Li H, Kim SJ, Lee YW, Kee C, Oh I (2011) Determination of the stoichiometry and critical oxygen tension in the production culture of bacterial cellulose using saccharified food wastes. Korean J Chem Eng 28(12):2306–2311
118. Watanabe K, Yamanaka S (1995) Effects of oxygen tension in the gaseous phase on production and physical properties of bacterial cellulose formed under static culture conditions, bioscience, biotechnology, and biochemistry. Biosci Biotech Biothem 59:65–68
119. Tantratian Sumate, Tammarate Pramote, Krusong Waravut, Bhattarakosol Pattarasinee, Phunsri Angkana (2005) Effect of dissolved oxygen on cellulose production by *Acetobacter* sp. J Sci Res Chula Univ 30(2):179–186
120. Naritomi Takaaki, Kouda Tohru, Yano Hisato, Yoshinaga Fumihiro (1998) Effect of ethanol on bacterial cellulose production in continuous culture from fructose. J Ferment Bioeng 85(6):598–603
121. Lu Z, Zhang Y, Chi Y, Xu N, Yao W, Sun B (2011) Effects of alcohols on bacterial cellulose production by *Acetobacter xylinum* 186. World J Microbiol Biotechnol 27(10):2281–2285
122. Yunoki Shunji, Osada Yoshiko, Takai Hiroyuki Kono Mitsuo (2004) Role of ethanol in improvement of bacterial cellulose production: analysis using 13C-labeled carbon sources. Food Sci Technol Res 10(3):307–313
123. Agustin YE, Padmawijaya KS (2018) Effect of acetic acid and ethanol as additives on bacterial cellulose production by *Acetobacter xylinum*. IOP Conf Ser Earth Environ Sci 209:012045
124. Zhao Hongwei, Xia Jian, Wang Jiming, Yan Xiaofei, Wang Cong, Lei Tingzhou, Xian Mo, Zhang Haibo (2018) Production of bacterial cellulose using polysaccharide fermentation wastewater as inexpensive nutrient sources. Biotechnol Biotechnol Equip 32(2):350–356. https://doi.org/10.1080/13102818.2017.1418673
125. Ishida Takehiko, Mitarai Makoto, Sugano Yasushi, Shoda Makoto (2003) Role of water-soluble polysaccharides in bacterial cellulose production. Biotechnol Bioeng 83(4):474–478
126. Hongmei Lu, Jia Qinghui, Chen Li, Zhang Liping (2016) Effect of organic acids on bacterial cellulose produced by *Acetobacter xylinum*. Res Rev J Microbiol Biotechnol 5(2):1–6
127. Zhou LL, Sun DP, Hu LY, Li YW, Yang JZ (2007) Effect of addition of sodium alginate on bacterial cellulose production by *Acetobacter xylinum*. J Ind Microbiol Biotechnol 34:483–489
128. Pokalwar SU (2011) Studies on cellulose production by using bacteria. PhD Thesis, Swami Ramanand Teerth Marathwada University, Nanded. http://hdl.handle.net/10603/210178
129. Zakaria J, Nazeri MA (2012) Optimization of bacterial cellulose production from pineapple waste: effect of temperature, pH and concentration. In: EnCon 2012, 5th Engineering Conference, "Engineering Towards Change - Empowering Green Solutions", 10–12 Jul 2012
130. Rangaswamy BE, Vanitha KP, Basavaraj S (2015) Hungund, microbial cellulose production from bacteria isolated from rotten fruit. Int J Polym Sci. https://doi.org/10.1155/2015/280784

131. Goh WN, Rosma A, Kaur B, Fazilah A, Karim AA, Bhat Rajeev (2012) Fermentation of black tea broth (Kombucha): I. Effects of sucrose concentration and fermentation time on the yield of microbial cellulose. Int Food Res J 19(1):109–117
132. Yamanaka S, Watanabe K (1994) Applications of bacterial cellulose in cellulosic polymers. In: Gilbert R (ed) Cellulosic polymers, blends and composites. Hansen Verlag, Munchen, pp 207–215
133. Yanti NA, Ahmad SW, Muhiddin NH (2018) Evaluation of inoculum size and fermentation period for bacterial cellulose production from sago liquid waste. IOP Conf Ser J Phys Conf Ser 1116:052076. https://doi.org/10.1088/1742-6596/1116/5/052076
134. Hungund BS, Gupta SG (2010) Improved production of bacterial cellulose from *Gluconacetobacter persimmonis* GH-2. J Microbial Biochem Technol 2(5):127–133
135. Gayathrya G, Gopalaswamy G (2014) Production and characterisation of microbial cellulosic 620 fibre from *Acetobacter xylinum*. J Fibre Text Res 39(1):93–96
136. Gunduz G, Kiziltas EE, Kiziltas A, Gencer A, Aydemir D, Asik N (2018) Production of bacterial cellulose fibers in the presence of effective microorganism. J Nat Fibers. https://doi.org/10.1080/15440478.2018.1428847
137. Raiszadeh-Jahromi Y, Rezazadeh-Bari M, Almasi H et al (2020) Optimization of bacterial cellulose production by *Komagataeibacter xylinus* PTCC 1734 in a low-cost medium using optimal combined design. J Food Sci Technol 57:2524–2533
138. Ghalachyan A (2018) Evaluation of consumer perceptions and acceptance of sustainable fashion products made of bacterial cellulose. Graduate theses and dissertations. 16583. https://lib.dr.iastate.edu/etd/16583
139. Solatorio N, Chong Liao C (2019) Synthesis of cellulose by *Acetobacter xylinum*: a comparison vegan leather to animal and imitation leather. Honors thesis, University of Wyoming
140. Ghalachyan A, Lee YA, Changhyun N (2016) Improving properties of bacterial cellulose by incorporating bio-based films.In: International Textile and Apparel Association (ITAA) Annual Conference Proceedings, p 133. https://lib.dr.iastate.edu/itaa_proceedings/2016/posters/13

Chapter 3
Characteristics of Bacterial Cellulose

Textile and Fashion Perspective

3.1 Introduction

Textile and fashion industry is suffering from serious pollution issues in every manufacturing activity from cotton cultivation to product disposal. Several methodologies have been proposed by various researchers to convert the process and products as a sustainable one. But due to the working nature and mass production capacity of industry, the possibilities of converting those research scale models are still in infant stage. Bacterial cellulose is one of such material which has lot of potential in textile and fashion sector. However, the properties and limitations of bacterial cellulose are less explored in this aspect. Out of various production methods discussed in the previous chapters, the methods that can produce bacterial cellulose in the form of continuous sheets (batch method, bioreactors, etc.) are more preferred in the case of textile and fashion applications. This chapter is aimed to analyse the different physical and chemical structures and properties of the bacterial cellulose developed in the 3D matrix or sheet form. The morphological analysis is used to reveal its structural arrangements of fibrils, size and shapes of the fibrils inside the matrix. The chemical properties of the bacterial cellulose were detailed with the structural and crystalline arrangement of the sheet using FTIR and XRD analysis. The thermal stability of the cellulose matrix is also evaluated using differential scanning calorimetry (DSC/TGA) analysis. The properties were analysed in comparison with the plant cellulose (cotton) in order to report the potentiality of the bacterial cellulose in textile and fashion application. The second section of the chapter elucidates the various post-processing methods needed or used in the bacterial cellulose to convert it as a fabric. In which, the effect of NaOH treatment, bleaching process to develop pure white material and dyeing characteristics with different colour medium to enhance the applications are detailed. The last and third part of the chapter details the various drying techniques and its influence on the functional properties of the raw and treated bacterial cellulose material.

© The Author(s), under exclusive license to Springer Nature Singapore Pte Ltd. 2021

S. S. Muthu and R. Rathinamoorthy, *Bacterial Cellulose*, Sustainable Textiles: Production, Processing, Manufacturing & Chemistry,
https://doi.org/10.1007/978-981-15-9581-3_3

3.2 Morphological Properties

The morphological structure of the bacterial cellulose is one of the main reasons for its unique and special properties. The size and structure of the cellulose fibre are associated with various parameters like incubation time, types of carbon source used and many other factors. However, the major structural variations in the fibre matrix are mainly associated with the type of fermentation method, namely static cultivation and agitated cultivation method. As discussed in the earlier chapter, the oxygen availability in the culture media throughout the incubation also has a potential impact on the structure formation. The above-mentioned methods have higher differences concerning the dissolved oxygen content in the culture media; hence, it is important to discuss the factor. Figure 3.1 represents the various production parameters that directly influence the different properties of bacterial cellulose, namely physical, mechanical, thermal and other properties.

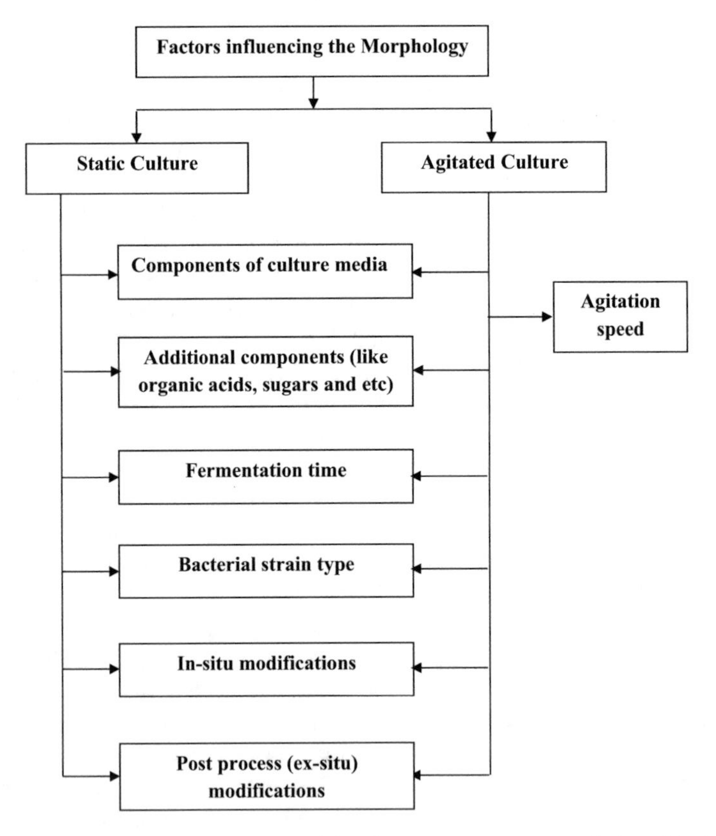

Fig. 3.1 Parameters influencing the bacterial cellulose structure or morphology [Authors own illustration]

During production, the bacterial cellulose obtained immediately after a long period of fermentation is used to be embedded with millions of micro-organisms along with the different medium components. As the production of cellulose is performed by the bacterial strain, this cannot be avoided. The wet, just fermented bacterial cellulose cannot be used as such for other applications. Hence, the produced cellulose is rinsed in the running water for a longer duration or immersed or treated with mild alkali to remove the above-said components. This will remove unwanted impurities from the bacterial cellulose pellicle and provide pure cellulose for the subsequent process. It is important to mention that this is an essential step, non-removal of the embedded bacteria (in either a dead or alive form) and media component in the cellulose itself will result in an unpleasant odour upon storage. This may further help to grow other micro-organisms like fungus on the surface when it is exposed to moisture. The internal morphological structure of the cellulose fibrils and the overall three-dimensional appearance will be exposed only after the pretreatment process.

In the static culture method, due to the undisturbed long fermentation, a continuous sheet-like cellulose fabric can be produced. This structure is one of the most explored aspects of bacterial cellulose. The appearance of bacterial cellulose produced in this method is reported in Fig. 3.2. The cellulose pellicle has a gelatinous structure, formed on the surface of the culture media. Various researchers tried to analyse its

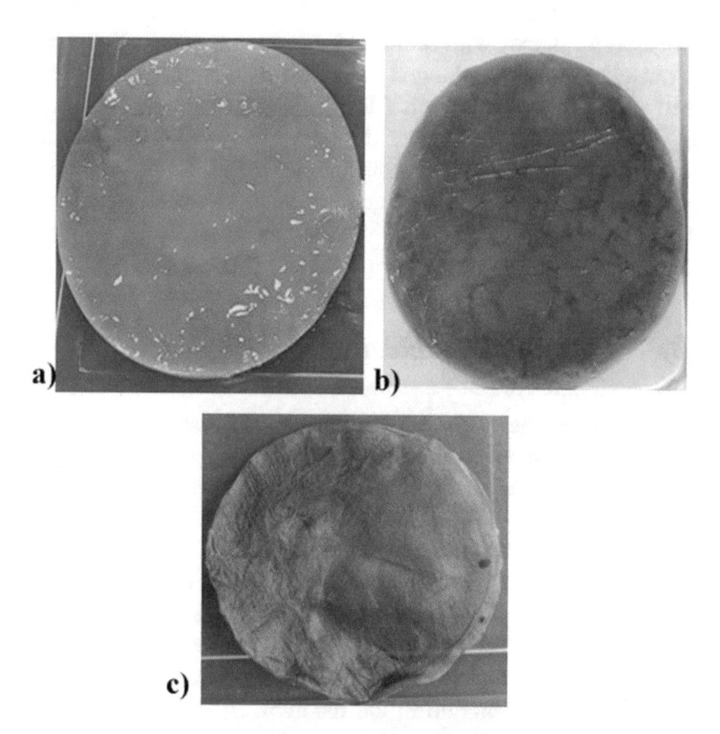

Fig. 3.2 **a** Untreated bacterial cellulose (wet), **b** washed bacterial cellulose (wet) and **c** dried bacterial cellulose from static culture (static method) [Authors own illustration]

Fig. 3.3 Bacterial cellulose produced from agitated culture (spherical shape) [1] [Reprinted with Permission]

morphology and identified that the individual fibre size is in the nano-size range. In an agitated medium, the cellulose is produced in the form of spherical or ellipsoidal or stellate or fibrous shapes with different sizes as reported in Fig. 3.3. Similarly, a different bacterial strain, production media, fermentation time, different carbon sources used and additives used will alter the structure of the bacterial cellulose. A different culture medium influences the microstructure of the developed bacterial cellulose by adjusting the degree of polymerisation, the crystallite size of the cellulose, crystallinity index and the percentage of the cellulose type (cellulose I and II).

3.2.1 Effect of In-Situ Modification

In-situ modification is one of the most common processes in the bacterial cellulose production to enhance cellulose production or to impart specific structural properties. Several research workers had performed in-situ modification by altering culture media, additives and other components as detailed in Chap. 2. The modification broadly affects cellulose in terms of yield, properties and functional aspects. However, a significant structural change is the fundamental technical reason for all the changes occurring in the bacterial cellulose. In a study, the researcher used oligosaccharide along with the common synthetic media. The researcher noted significant differences in the structural compactness and the increase of fibre thickness due to the addition of oligosaccharide in the medium. By these changes, the overall porosity of the bacterial cellulose matrix reduced significantly. A similar effect can also be obtained by increasing the fermentation duration. The longer fermentation time increases the amount of cellulose production in individual cells, and it increases the density of the cellulose in the mat and so the thickness [2]. The findings revealed that the addition of different concentrations of oligosaccharide (1, 2 and 4%) significantly reduced the pore volume and total surface area. Out of all the samples, the native cellulose showed a higher surface area of 178 m^2/g and the addition of oligosaccharide reduced it to 168, 135 and 104 m^2/g, respectively, for 1, 2 and 4% of the oligosaccharide. Similarly, the pore volume also noted high with respect to native cellulose (0.505 cc/g) and reduced with successive addition of oligosaccharide [2]. In the case of the addition of a different volume of carboxymethyl cellulose in the bacterial cellulose, growth increased the adherence of the microfibrils in the morphology. The structure becomes more compact than the native cellulose, and the thickness of the fibril ribbon increased significantly [3].

Mudrika Khandelwal et al. analysed the effect of carboxymethyl cellulose (CMC), polyethylene glycol (PEG), Calcofluor (CF) and nalidixic acid (NA) as an additive in the bacterial cellulose production and analysed the morphology. The findings revealed that the addition of different components reacted differently with the bacterial cellulose structure. In common, branching of fibrils is noted in all the materials used for the modification with different distribution lengths for a different component. When compared to the native cellulose, the CMC-added bacterial cellulose sample fibre ribbons showed a lower width. In the case of PEG, no difference is noted with the fibril structure. The researchers also noted an increment in the surface area of the PEG-treated bacterial cellulose over other compositions. These findings confirmed that based on the filler material or additives used in the in-situ bacterial cellulose preparation, the porosity of the material can be controlled effectively [4]. Another research evaluated the use of non-water-soluble poly-3-hydroxybutyrate in the in-situ modification and its effects on the structural changes by comparing it with water-soluble polymers like hydroxypropyl methylcellulose and Tween 80. The results reported a thickness variation in the bacterial cellulose. The poly-3-hydroxybutyrate-added bacteria produced a thinner fibre structure than the hydrophilic additives. In specific, Tween 80 generated a thicker fibre. This is attributed to the gluconic acid generation in cellulose growth. In particular, the poly-3-hydroxybutyrate was largely coated on the surface of the developed pellicle. Further, the presence of poly-3-hydroxybutyrate in between the adjacent layers of cellulose fibrils is also noted. The analysis further showed a deposition on the top surface of the pellicle in the form of powder. Not many changes are noted in the structure; however, there is a great reduction in the mechanical properties like tensile strength and modulus due to the changes in the crystallinity of the bacterial cellulose produced in situ with poly-3-hydroxybutyrate [5].

3.2.2 *Effect of Ex-Situ Modification*

The ex-situ modification process is generally performed after the cultivation of the bacterial cellulose. The developed bacterial cellulose is afterwards treated or impregnated or surface grafted or functionalised with different chemicals based on the requirements. The functionalisation process is commonly performed by solubilising functional chemicals like chitosan, polyvinyl alcohol, etc., in suitable solvents. Hence, the microstructure of the bacterial cellulose allows it to penetrate inter- and intracellulose spaces and modifies the structural properties of the native (original) bacterial cellulose. In the ex-situ process, the research work sometimes showed a difference in the fibre diameter. However, based on the concentration of the material used, the deposition of the chemical can be observed on the surface or a full surface coating on the fibre surface can be observed. In an analysis, researchers applied silver nanoparticle with cerium nitrate to functionalise the bacterial cellulose. The results showed a complete fibre coating and a slight increment in the fibre diameter. The scanning electron microscope analysis clearly showed that it is deposited on the surface and the original three-dimensional porous structure of the material maintained without alteration. The images of the treated bacterial cellulose are provided in Fig. 3.4 for better understanding [6].

The application of chemicals may also lead to more aggregation inside the structure. Based on the diameter of the particles, the deposition pattern may get changed. In the ex-situ analysis study, bacterial cellulose is treated with silver nanoparticles to provide antimicrobial properties to the bacterial cellulose for wound dressing application. The morphological analysis revealed that the treatment did not affect the porosity. However, the silver nanoparticles are scattered all around the fibre as either individual particles or aggregates inside the structure [7]. In a similar treatment, based on the material used sometimes several morphological changes can be obtained. While applying aminoalkyl silane groups for functionalising, the porous and three-dimensional structures become denser due to the filling of the functional

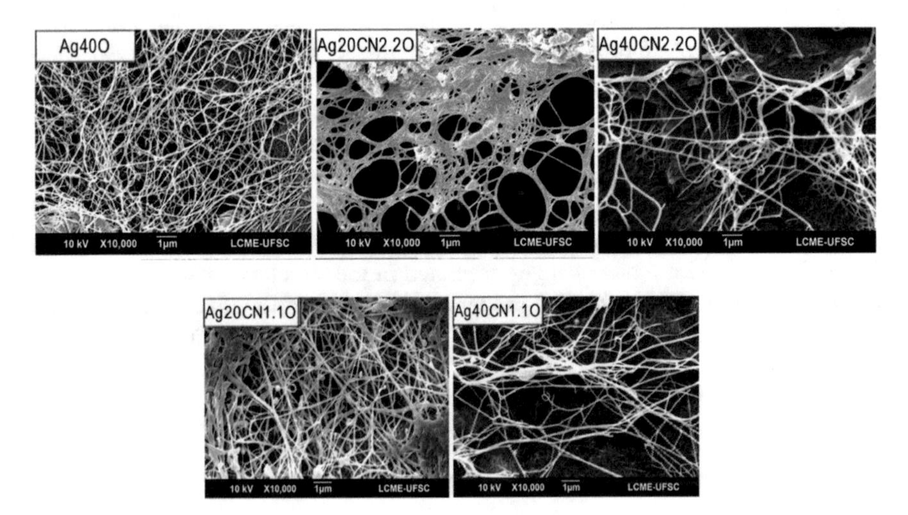

Fig. 3.4 Ex-situ modified bacterial cellulose showed the coating and deposition of applied function material at different concentration [6] [Reprinted with Permission]

compound in the structure. Further, the cross-sectional analysis revealed that the modification process created a lamellar structure, due to the repulsive forces created between the fibres by the aminoalkyl silane groups [8]. Use of polyvinyl alcohol, in the composite preparation with bacterial cellulose, had a significant influence on the surface topography of the cellulose matrix. The treatment initially filled the interfibre pores and made the structure more compact than native bacterial cellulose. Further, the treatment reduced the surface roughness of the bacterial cellulose significantly by depositing on the fibres. This property will alter the moisture handling behaviour of the bacterial cellulose. The author suggested that the film-forming capacity of the polyvinyl alcohol was the main reason as it coats the fibres on the top surfaces, and it subsequently reduces the unevenness on the surface [9].

3.3 Chemical Properties

Chemical properties of the bacterial cellulose were commonly analysed by Fourier transform infrared spectroscopy (FTIR) and X-ray diffraction (XRD) analysis. This reveals the internal chemical bonds and crystalline natures and cellulose types present in the developed bacterial cellulose. It is essential to estimate the bacterial cellulose properties which differ in a large manner concerning the method it is produced or post-treatment imparted.

3.3.1 FTIR Analysis

Bacterial cellulose is known for its highly pure cellulose content over plant cellulose due to the biological origin. The FTIR spectrum shows intensity peaks as a representative of cellulose. On the analysis of bacterial cellulose produced, researcher reported functional peak at 3350 cm^{-1}, 1500–1235 cm^{-1} region by representing O–H stretching of cellulose. The C–H stretching is noted in the region of 2690 cm^{-1}. Due to the scissoring action of CH_2 bonds, the researcher also reported a strong bond at 1424.18 cm^{-1} and broadband at 891 cm^{-1} [1]. The typical curve reported by Rathinamoorthy et al. is provided in Fig. 3.5, representing the native bacterial cellulose without any purification process [10].

Several representative peaks reported by other researchers to identify the bacterial cellulose are provided in Table 3.1 for better understating.

The effect of chemical treatment like alkali and bleaching agents on the structural properties of the bacterial cellulose was evaluated by Han et al. [18]. In comparison with the native bacterial cellulose spectrum with bleached one, they had reported an amide II absorption peak at 1533 cm^{-1} in the untreated bacterial cellulose. This represents the presence of amides, N–H stretching of proteins and amino acids on the surface from bacteria. The chemically treated bacterial cellulose mats showed the disappearance of 1533 cm^{-1} peak, by representing the removal of impurities.

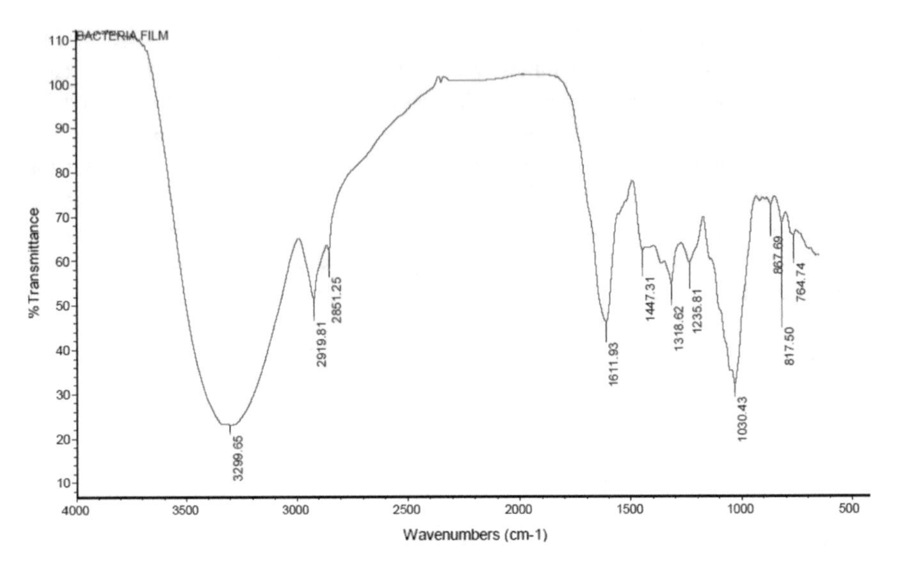

Fig. 3.5 FT-IR spectrum of native bacterial cellulose [10] [Reprinted with Permission]

Instead, they had noted a solid representation of cellulose I at 1422 cm^{-1} due to the CH2 bending or O–H in-plane bending. At the same time, the researchers also reported the presence of standard cellulose I peaks at 1107–1109 cm^{-1} by mentioning the CH_2 symmetrical bending and C–O bond stretching. Similarly, the peak at 1055–1031 cm^{-1} is indicating the (C–O of the C–OH) carbohydrates of cellulose I component [18]. After chemical treatment, the bacterial cellulose materials are most commonly functionalised with in-situ and ex-situ modification. The developed composites usually provide some morphological changes as discussed earlier. However, in the case of chemical structure, specifically with FTIR analysis, the researchers can always find the chemical groups that represent both the components, namely bacterial cellulose and functional chemical.

Sudipto Pal et al. evaluated the chemical changes in the composite developed using silver nanoparticle and bacterial cellulose. The FTIR spectrum showed the peaks to represent the native bacterial cellulose (at 3200–3400, 2800–2900 and 1200–900 cm^{-1}) and also to identify silver nanoparticles (1387 cm^{-1}). The spectrum is provided in Fig. 3.6 for comparison purposes [7]. Similarly, the addition of lauryl gallate oligomers through the ex-situ method showed a similar effect on the structure of the bacterial cellulose. The research represented the developed composite showed all standard peaks of native bacterial cellulose along with the functional material incorporated. The intensity peaks at 2940–2893 cm^{-1} showed a small change due to the incorporation of lauryl gallate. The researcher also noted a few changes in the spectrum due to the addition of lauryl gallate. They reported carbonyl stretch by esters at 1710 cm^{-1}, and there was a change noted in the amine peak (1590 cm^{-1}). The reduction in water quantity peak at 3300 cm^{-1} is also largely attributed to the addition of lauryl gallate oligomers [19].

Table 3.1 FTIR peak representations to indicate the of bacterial cellulose

S. no.	Peak wavelength (cm^{-1})	Chemical bond	Representation	Reference
1.	3495	OH stretching	Cellulose II	Moharram and Mahmoud [11]
2.	3420	Stretching of N–H free	–	Oh et al. [12]
3.	3300	OH stretching	Inter- and intravibration of cellulose I	Wonga et al. [13]
4.	3278	OH stretching	Cellulose I vibration	Moharram and Mahmoud [11]
5.	3240	H-bonded O–H	Represents the presence of cellulose Iα	Focher et al. [14]
6.	2912	CH stretching	Amorphous regions of cellulose	Rathinamoorthy et al. [10]
7.	2860	CH stretching	–	Oh et al. [12]
8.	1730–1735	C=O stretching	Represents proteins and lipids	Movasaghi et al. [15]
9.	1611	COOH	Carboxylic acid groups	Rathinamoorthy et al. [10]
10.	1630	H–O–H bond	Presence of moisture	Rathinamoorthy et al. [10]
11.	1425–1435	In-plane bending of H–C=H and O–C=H	–	Yassine et al. [16]
12.	1318	–	Glucose carbonyl cellulose	Yassine et al. [16]
13.	1146–1160	Asymmetric stretching of C–O–C, CH deformation	–	Kacurakova et al. [17]
14.	1031.4	C–O–C	–	Yassine et al. [16]
15.	870–900	CH– out-of-plane bending		Kacurakova et al. [17]

3.3.2 XRD Analysis

XRD analysis identifies the crystallinity of the material. Bacterial cellulose generally produces two forms of cellulose represented as cellulose I and cellulose II. In cellulose I structure, the β-1,4 glucan chains are arranged parallel by default. Whereas in cellulose II, the molecular chain arranged randomly. Both these structures are more common in all types of cellulose from plants, wood and micro-organism. Out of

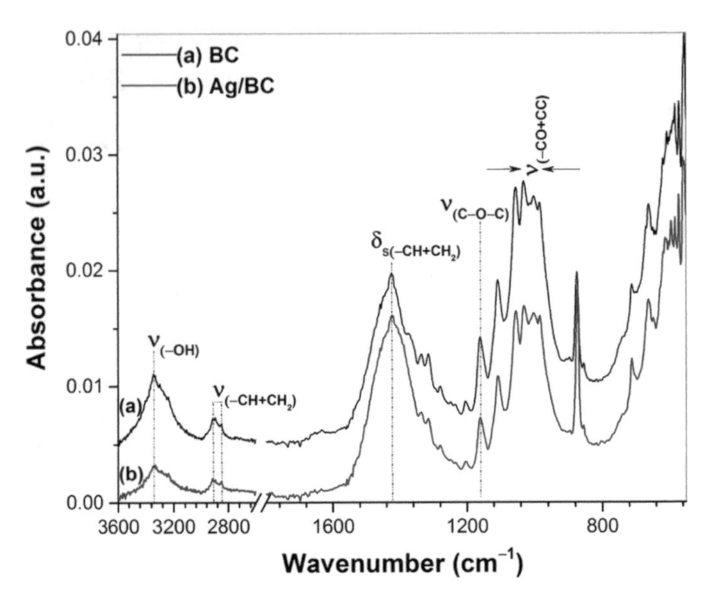

Fig. 3.6 FTIR spectra of the native bacterial cellulose (BC) and silver nanoparticle-deposited (Ag/BC) nanocomposite [7] [Reprinted with Permission]

these two structures, cellulose II represents a highly stable form due to the presence of more number of hydrogen bonds internally [20]. In common bacterial cellulose, peaks appeared in two major regions of 14, 16.5 and 22° in the XRD pattern, which typically represents the cellulose Iα and Iβ phase. These peaks are corresponding to the (110), (110) and (200) diffraction planes, respectively. These two types of crystalline phases are the major component of the native bacterial cellulose [21]. Other than this, the XRD analysis reveals the crystallinity percentage of bacterial cellulose, which is one of the important factors for the durable thermal and mechanical properties. The crystallinity of the produced cellulose is mainly based on the type of medium used. Bacterial species generally produce higher crystalline cellulose in the HS medium, the standard synthetic medium. A crystallinity of 93% is reported by Kulkarni et al. [22].

On the cellulose produced from different feedstock materials, the crystallinity percentage is noted differently. A research work evaluated the effect of different industrial wastes or by-products as a feedstock for bacterial cellulose production. They analysed the crude glycerol combined with glucose-free HS medium (BC1), crude glycerol combined with solid fermentation hydrolysate medium (BC2) from confectionery industry waste streams, HS medium (BC3) (biodiesel industry by-products) and flour-rich waste hydrolysate (BC4) medium. The results reported a higher crystallinity percentage of 88 and 89% for bacterial cellulose produced from mediums 1 and 4. The other mediums 2 and 3 showed a crystallinity of 74% and 81%, respectively. The XRD spectrum of bacterial cellulose produced from different sources is provided in Fig. 3.7 [23].

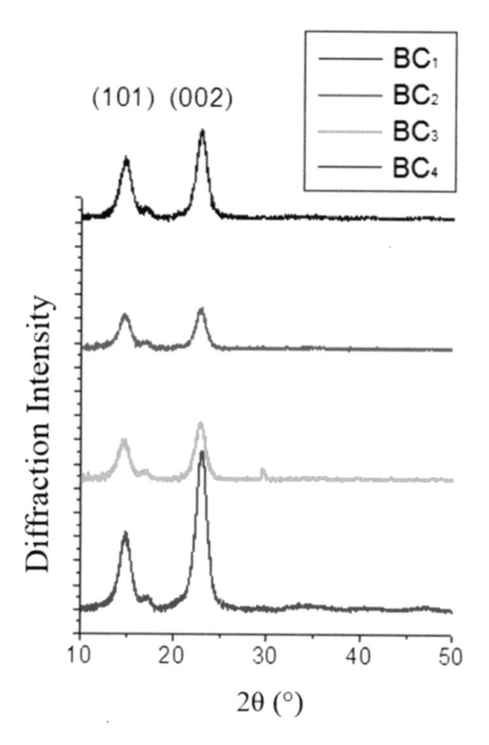

Fig. 3.7 XRD spectrum of bacterial celluloses produced from different industrial wastes [23] [Reprinted with Permission]

The purification process like alkali treatment and bleaching had a considerable impact on the molecular structure. The alkali and bleach treatment reduces the impurities and so realigns the cellulose fibrils in a better way. The report provided by researchers showed that the subsequent process increased the crystallinity of the bacterial cellulose. The crystallinity of the boiled cellulose is initially noted as 53.3%, after the alkaline treatment it is increased to 73%, and after bleaching it became 80%. This is mainly because of the size reduction in the cellulose crystallite due to the alkali treatment. The results mentioned that the crystallite size of the cellulose reduced from 76.4 to 63.7 A° [18]. In the case of the ex-situ treatment, bacterial cellulose treated with lauryl gallate monomer showed an increment in the crystallinity from 73 to 79%. This is due to the higher alignment of microfibril inside the structure. This in turn reduces the moisture absorbency of the material [20]. The addition of silver nanoparticle to the bacterial cellulose showed diffraction peaks at 2θ, 38.1, 44.2, 64.4 and 76.7° due to the (111), (200), (220) and (311) reflections which confirm the presence of both bacterial cellulose and silver nanoparticle. The diffraction peaks also represent the presence of silver nanoparticles. However, the crystallinity of the bacterial cellulose reduced after the incorporation of silver nanoparticles from 86.8 to 81.4% [7]. The effect of chitosan and glycerol on the bacterial cellulose reduced the crystallinity percentage. The research results showed that the ex-situ composite has reduced crystallinity of 47% in the case of glycerol addition and in the case of chitosan it is 50.15%, where in the native bacterial cellulose it is noted as 73.65%. The reduction in the property might be attributed to the amorphous nature of the materials used in the composite preparation [24].

3.4 Thermal Properties

Thermo-gravimetric analysis and differential scanning calorimetry are the major types of analysis performed to understand the thermal behaviour of bacterial cellulose or any other material. In thermo-gravimetric analysis, the thermal property of the bacterial cellulose is analysed by measuring the changes in the mass on the application of heat. The analysis reveals the weight loss behaviour and thermal degradation ability of the material. The differential scanning calorimetry method is to evaluate the glass transition temperature and isothermal crystallisation characteristics of the material. The thermal degradation behaviour of any material is influenced by the molecular weight, crystallinity and orientation of the material.

3.4.1 Thermo-gravimetric Analysis

In this method, the material is subjected to a continuous heating process and the mass changes were noted against the schematic heat changes. It is a common method for all the materials, and it evaluates the mass change due to moisture loss and temperature-dependent degradation. This method provides a better understanding of the thermal decomposition nature of the material. In a raw and native bacterial cellulose analysis, the researcher reported a multistep degradation of the bacterial cellulose. Initially, there was degradation due to moisture loss and by low molecular elements followed by temperature-oriented degradation. This step represents the decomposition of the bacterial molecular structure. The complete decomposition of bacterial cellulose has occurred around 600 °C. The multistage degradation curve of bacterial cellulose is provided in Fig. 3.8 [10].

The chemical treatments like sodium hydroxide and other alkaline treatments are more common in the case of bacterial cellulose. To remove the culture media content and also to remove other impurities like dead bacteria from the cellulose non-woven matrix, the process is usually performed. The treatment also has an impact on the thermal properties. A bacterial cellulose sample treated with various alkalis like K_2CO_3, Na_2CO_3, KOH and NaOH were analysed for its thermal degradation properties. The results indicated that the treatment process increased the degradation temperature from 150 to 275 °C. The mass loss percentage was considerably reduced from 26% to 6–7% in the subsequent temperature. The higher mass loss in the native cellulose is associated with the presence of impurities. The alkali treatment reduced the impurities, and so the mass loss percentage is reduced in the alkali-treated sample.

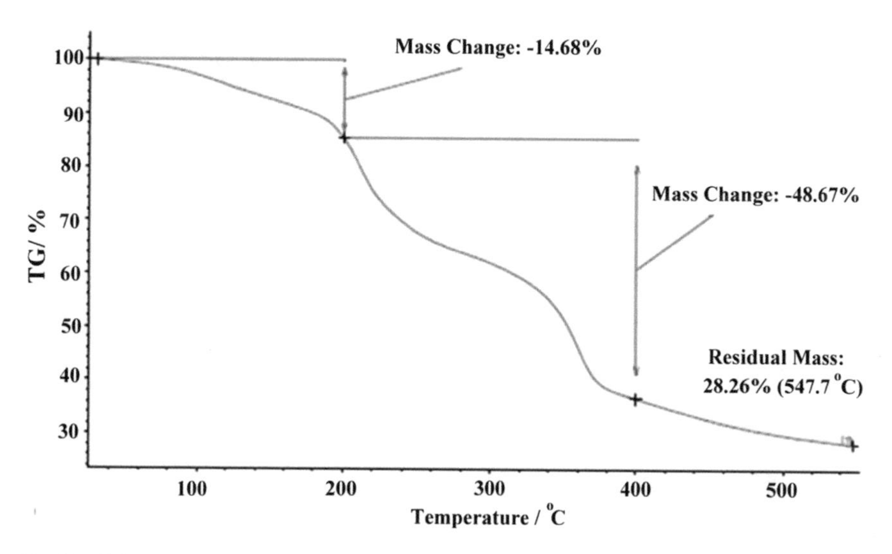

Fig. 3.8 TGA curve of the raw bacterial cellulose [10] [Reprinted with Permission]

Concerning the degradation temperature, the researcher reported two different peaks at 158 and 312 °C by representing the degradation of proteinaceous matter and cellulose, respectively. The alkali-treated samples showed a single cellulose degradation peak at 350 °C due to the removal of proteinaceous matters in the previous stage. The chemical treatment removes low molecular weight substances and orients the cellulose fibrils in a better manner [25]. The different production methods may alter the physical form of the bacterial cellulose. However, there is no much difference noted in the thermal degradation properties of the bacterial cellulose. The bacterial cellulose produced using one of the production methods called bioreactors did not have any influence on the thermal properties. The study reported the two degradation peaks at 265 and 445 °C by representing a similar curve related to the native bacterial cellulose [26].

Chitosan is one of the commonly used functional materials along with bacterial cellulose to develop application-specific composites. The addition of chitosan either in situ or ex situ alters the thermal stability of the material strongly. The addition of different concentrations of chitosan in the cellulose medium had a significant effect on the degradation. The degradation behaviour of the various concentrations of chitosan-added bacterial cellulose for the temperature range are provided in Fig. 3.9. The results indicated that the addition of chitosan reduced the degradation temperature of the chitosan bacterial cellulose composite. Though all the composites showed a classical two degradation peaks, the temperature range is reduced to 300 and 345 °C for an extended time than pure bacterial cellulose. The reduction in the temperature of the composite might be attributed to the inclusion of a lower-temperature chitosan component [27]. Other researchers treated polyphosphate solutions at different molar ratios with bacterial cellulose by ex-situ analysis. The findings reported that the composite possessed two peaks. The initial mass loss on the composite is up to 200 °C

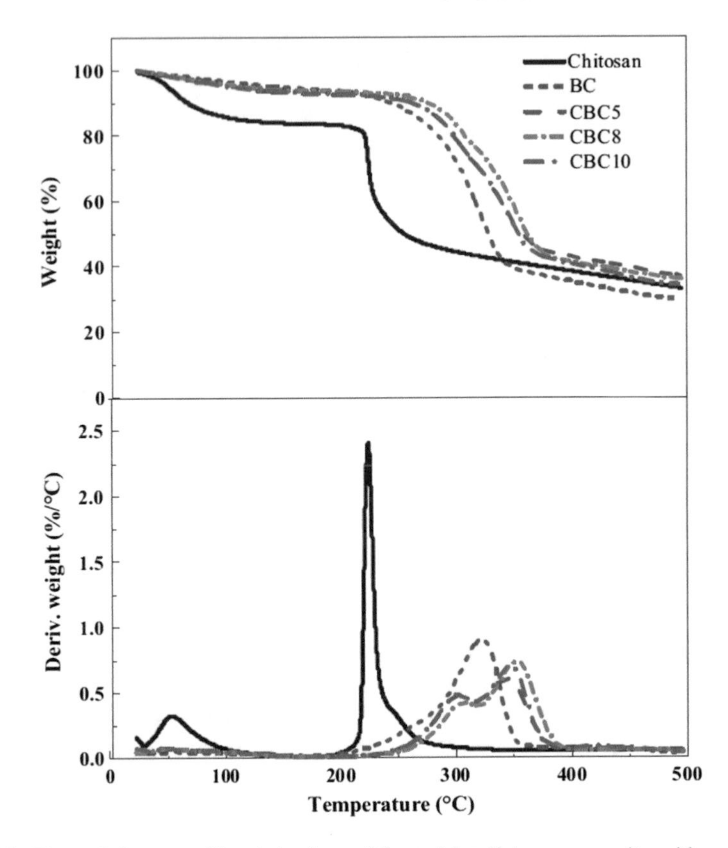

Fig. 3.9 Thermal decomposition behaviour of bacterial cellulose composite with a different chitosan percentage [27] [Reprinted with Permission]

due to the moisture loss of the composite. From there to 400 °C, the second loss is due to the thermal degradation of the composite. The addition of polyphosphate increased the water content and so higher weight loss noted. The composite prepared with 20% of polyphosphate showed a similar behaviour of the native bacterial cellulose. However, the addition of polyphosphate significantly altered the glass transition temperature of the composite [28].

3.4.2 Differential Scanning Calorimetry

The scanning calorimetry study result represents the glass transition temperature of the bacterial cellulose material. The typical curve of the bacterial cellulose is provided in Fig. 3.10. The findings reported an endothermic peak at 121.3 °C. The glass transition temperature of the native bacterial cellulose is noted as 165 °C (midpoint)

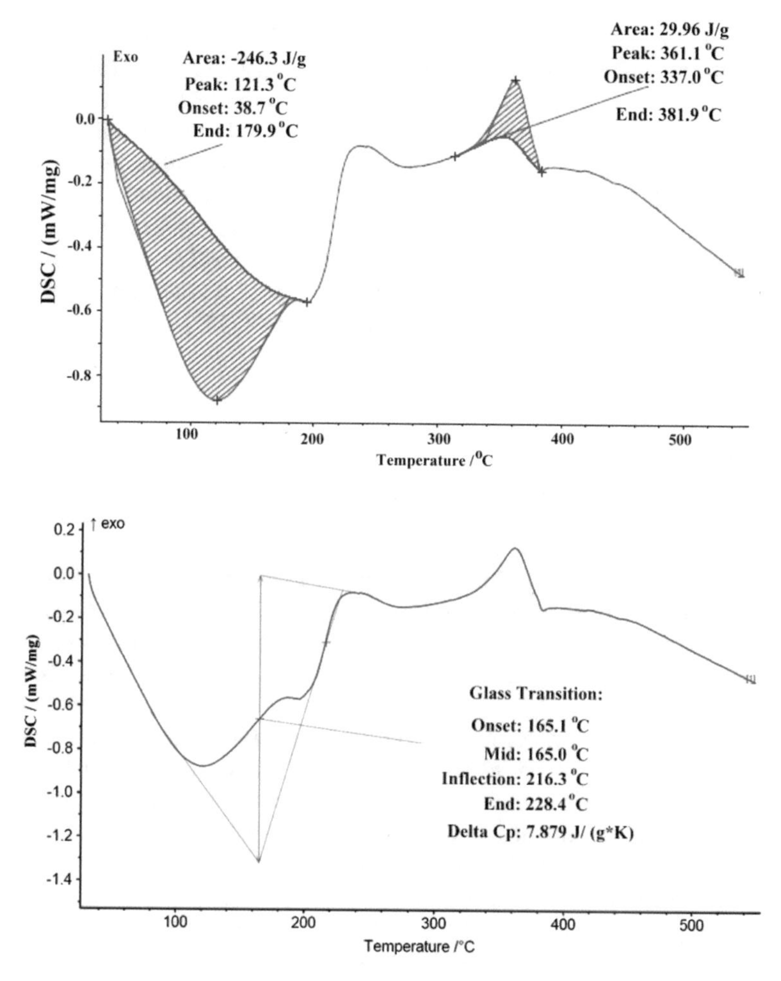

Fig. 3.10 Differential scanning calorimetry spectra of native bacterial cellulose film [10] [Reprinted with permission]

in the endothermic peak (Fig. 3.10). The higher glass transition temperature of the native bacterial cellulose is mainly due to its higher crystallinity [10].

In this analysis, the first endothermic curve is a representation of the crystalline melting temperature of the polymer. The degradation peak (exothermic peak) around 300 °C of the native bacterial cellulose is the typical representation of partial pyrolysis of carbonyl and carboxylic bonds from anhydrous glucose [28]. The thermal degradation steps start from 80 °C, and the melting of the crystalline phase happens at the range of 80–140 °C in the bacterial cellulose. The native bacterial cellulose membrane can withstand up to 343 °C without degradation, and after that, the polymorphic transition is followed by solid-phase melting. This higher thermal stability of the bacterial cellulose is mainly attributed to the higher crystallinity percentage.

The general glass transition temperature noted in a wide range from 40 to 160 °C is based on the type of culture medium and production methods adapted. The melting temperature of the bacterial cellulose is noted in the range of 85 °C [1]. The results of the previous research works reported that the in-situ and ex-situ modifications had a significant impact on the thermal degradation and glass transition temperature. In a study reported by Barud et al. after modification of bacterial cellulose structure by acetylation, the composite showed relatively higher stability on temperature. The results showed a glass transition temperature in the range of 174–184 °C. Similarly, higher crystallisation temperature and melting temperature are also noted for the developed composite, respectively, as 299 °C and 400 °C [29].

3.5 Mechanical Properties

Mechanical properties of the bacterial cellulose are mainly dependent on the chemical and morphological properties. Higher fibre orientation and crystallinity percentage provide a stronger structure that can withstand more mechanical stress. As the crystallinity differs based on the source and types of production methods, the mechanical properties also vary accordingly. In the same way, post-chemical treatments and ex-situ modification also either increase or lower the mechanical properties based on the type of chemicals used for composite preparation. The tensile strength of the raw bacterial cellulose is noted as 13.83 kgf/cm^2. The researcher noted the value of bacterial cellulose non-woven material with cotton fabric (with similar thickness), and the results noted are comparable. The elongation value of the bacterial cellulose is reported as 10 mm, and in the case of cotton, it is noted as 21.4 mm. The report mentioned that the higher tensile strength is mainly due to the higher crystallinity and due to the use of green tea medium as a nutrient source in cultivation [10]. In a similar study, the researcher developed bacterial cellulose fabric with tea as a nitrogen source and sugar as a carbon source. The tensile strength value of the bacterial cellulose is noted 177.6 $\pm$ 8.88 N/mm^2 which is far higher than the top grain leather of the same thickness (71.74 $\pm$ 3.59 N/mm^2). Higher crystallinity of the bacterial cellulose was noted as the main reason for its higher strength. However, for the leather application considered, the thickness of the bacterial cellulose at dry state is lesser than leather. In the case of elongation percentage, the top grain leather showed a higher percentage (30%) than the bacterial cellulose (14%) [30]. Other researchers who developed bacterial cellulose using *Gluconacetobacter hansenii* NCIM 2529 in HS medium reported tensile strength of 14.4 Mpa and an elongation of 97.3% for agitated cultured cellulose. The research reported that the presence of more water molecules in the structure might have helped the fibre slide easily, due to its random oriented in-plan structure. Another one of the important parameters the researcher reported is Young's modulus. It represents the hardness of the material under compression stress. Due to the higher crystallinity of the material, Young's modulus is noted as 15.71 N/mm^2 [1]. In a comparative study of bacterial cellulose produced from different bacterial strains, namely *Acetobacter xylinum* and

Komgataeibacter saccharovorans, the mechanical property results mentioned that the *K. saccharovorans*-based bacterial cellulose showed a higher tensile strength in the dry and wet state. This was mainly due to the higher degree of polymerisation of this bacterial cellulose. But in the case of Young's modulus value, the *A. xylinum*-based bacterial cellulose showed a higher value in both dry and wet states [31].

The mechanical property of the bacterial cellulose is increased by the two-step purification process by NaOH and NaOCl. The purification process removes the impurities present in the cultured bacterial cellulose. The removal of proteinaceous material and culture media allows the nanocellulose in the bacterial cellulose structure to align more perfectly than the native one. Hence, this process increases the crystallinity of the bacterial cellulose after every purification step. This change ultimately reflects in the mechanical properties of the produced cellulose. The results reported that the tensile strength of the native cellulose is noted as 88.9 Mpa and it increased to 139, 207 Mpa for single-step purified and two-step purified bacterial cellulose. A similar increment is also noted with Young's modulus from 7.6 Gpa for native cellulose to 14 and 18 Gpa for a single step and two steps treated [32]. The ex-situ modification of bacterial cellulose has a higher amount of influence on the mechanical property. For instance, the composite developed from glycerol, a plasticiser, showed a lower tensile strength of 15.60 Mpa over native bacterial cellulose (22.48 Mpa). This is mainly attributed due to the plasticising effect of glycerol in the microstructure of bacterial cellulose that caused higher slippage in structure. This can be confirmed by the higher elongation percentage (28.12%) than the native cellulose (22.18%). The use of other functional materials into the ex-situ method also showed a similar modification in the bacterial cellulose composite based on the characteristics of the functional material. The use of chitosan, as an antibacterial material for wound dressing along with bacterial cellulose, greatly reduced the mechanical property.

The addition of chitosan in the bacterial cellulose reduced the tensile strength to 17.1 Mpa from 22.48 Mpa. Chitosan is a hydrophilic and amorphous material than bacterial cellulose; hence, the addition of chitosan reduced the tensile strength and there is an increment in elongation [24]. The addition of a different percentage of softener worked differently with bacterial cellulose and also on the tensile properties. The addition of softener at a different percentage increased the tensile strength up to 1%, and then it reduced for further increment up to 50%. However, the elongation percentage of bacterial cellulose–softener composite increased slightly due to the softener addition. Similar to that of softener, hydrophobic agents also showed a similar trend with the bacterial cellulose. After the addition of 1%, there was an increment in the tensile strength and the reduction is noted up to 50%, but at 50% slight increment observed with the bacterial cellulose composite [33].

3.6 Moisture-Related Properties

Moisture-related properties are very important for several applications. Especially in the apparel and textile industries, the moisture properties of the material play a vital role. Most of the research works are performed on the water holding capacity and water absorption capacity in the wet state. Apart from its higher crystallinity, the structural porosity and higher surface area are the two main aspects of the bacterial cellulose that helps in higher moisture properties. The generally proposed mechanism is that the absorbed water resides inside the cellulose fibre and forms hydrogen bonds with it [34]. Based on the production methods, fibril arrangements, surface area and porosity cellulose showed different water holding capacity percentages from 100 to 200 times. The types of culture media compositions, other additives added in the media and the bacterial strain are the important parameters that have a direct influence on the produced fibre structure. Based on the crystalline and amorphous percentage of the bacterial cellulose, the absorption capacity differs. The water molecules cannot enter into the crystalline region; hence, the absorption capacity depends on it. Next to that, the larger surface area and pore volume are the other factors that increase the water absorption by physical entrapment and hydroxyl links [2].

3.6.1 Water Holding Capacity/Water Absorption/Swelling/Water Retention Capacity

The bacterial cellulose can accommodate the water molecule inside the structure. The ability of the material can be measured by analysing the weight increment of the sample by comparing it with its dry weight. The parameter becomes crucial in applications like wound dressing; where the material is expected to absorb continuous wound exudates. The absorption capacity differs mainly based on the morphological structure of the bacterial cellulose and with the treatment methods. Patricia Cazón et al. reported the two-stage interaction mechanism of the moisture or water molecule and cellulose in their isotherm analysis. At the first stage, the free binding sites in the cellulose chain react with the water molecule through hydroxyl groups. This interaction forms a monolayer with water molecules in the structure. In the second stage, the formation of multiple layers occurs with the partially interacted links. This increases the volume of the water inside the cellulose structure by occupying the free spaces and pores [35].

In bacterial cellulose, the internal fibril structures are developed linearly by glucan units that are linked by β-1,4 glycosidic bonds. Once the water entered into the structure, it establishes inter- and intramolecular hydrogen bonds between the adjacent glucan units. This mechanism is provided in Fig. 3.11 as reported by the previous researcher [36]. The water holding capacity of the bacterial cellulose was evaluated by different methods, namely the shake method, rapid transfer method, vertical drain method, horizontal drain method, wipe method and vacuum method. The finding

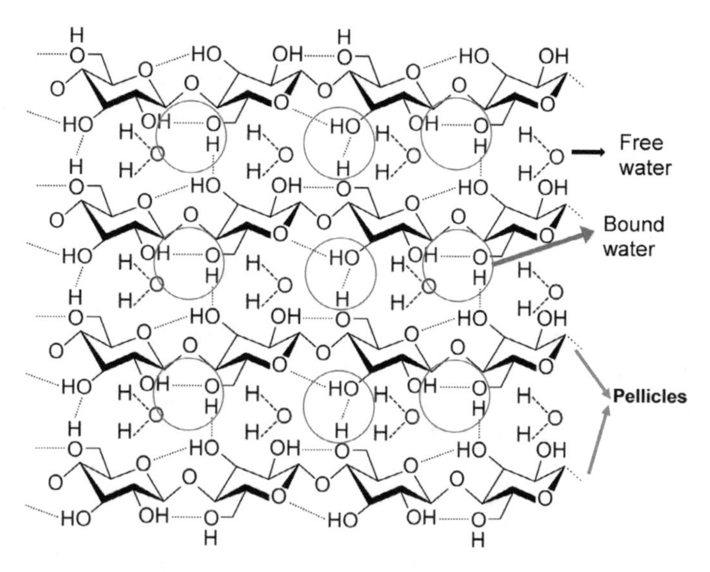

Fig. 3.11 Schematic of the molecular structure of bacterial cellulose and their bound and free water [36] [Reprinted with Permission]

revealed that the vacuum drain method is the most effective method to obtain a stable water holding capacity with a lower standard deviation than other methods. The vacuum method holds 148 g of water per gram of bacterial cellulose. This is the lowest water holding capacity among all the methods, and the maximum holding capacity of 291 g/g of cellulose is noted for the shake method. Longer duration and application of vacuum reduce the edge or surface saturation, edge dipping and other issues noted in the different methods [37].

Other researchers produced bacterial cellulose using an agitated culture method and analysed its water holding capacity. The water absorption capacity of the native bacterial cellulose is noted as 400%. The study revealed that the physical three-dimensional porous structures created a capillary force to suck the water molecule into it. This is one of the main reasons for its higher water holding capacity over hydrophilic nature. Similarly, the excessive hydrogen bonding between the reticulated cellulose fibril networks is also reported as one of the main reasons for higher water holding capacity. Generally, the water holding capacity of the bacterial cellulose obtained from the agitated culture is higher than the static culture fermented bacterial cellulose [1]. In a comparative study, the native bacterial cellulose is analysed with different concentrations of oligosaccharide in-situ modification. The addition of different concentrations of oligosaccharide extensively reduced the water holding capacity of the bacterial cellulose. The increment in the percentage of additional material in the medium increases the fibre diameter and reduction in pore volume in the matrix. Thus, it reduces the water holding capacity of the cellulose. The surface area and pore volume of the cellulose matrix are the important parameters to decide the water holding capacity. However, the results were different with the ex-situ sample. The native cellulose was immersed with the liquid chitosan solution and analysed for their water holding capacity. Chitosan, a functional material, is added mainly to impart antibacterial property in the bacterial cellulose. Similar to other additives, the addition of chitosan also reduced the interspace and porosity. However, due to the hydrophilic nature of the chitosan, the water holding capacity of the bacterial cellulose–chitosan composite is noted higher than the pure bacterial cellulose [2].

In the case of swelling measurement, upon immersing the bacterial cellulose in the deionised water for 24 h caused the material to absorb and swell 60 times its dry weight. This is far higher than a normal hydrocolloid film. At the same time, ex-situ modification of bacterial cellulose with chitosan and the bacterial cellulose–chitosan composite showed a very low swelling (nearly half of the native) ratio in contrast to the water absorption and holding capacity [38]. All the tests mentioned mostly evaluated the water holding capacity at a wet state. However, in the case of apparel, the rehydration properties are important. It represents the absorption of the liquid water after drying the bacterial cellulose. A study reported the absorbency of the bacterial cellulose was 320% of its dry weight and this is very higher than any textile material and polymer. The researcher suggested again that the higher surface area and pore volume are the main reasons for the same [10].

3.6.2 Water Vapour Transmission

The water vapour transmission is the ability of textile material representing an indirect measure of porosity of the material. It measures the permeability of the water vapour molecule into the material. This test is mainly performed for the textile material which is used in apparel which is in direct contact with skin. The objective is to evaluate the ability of the fabric to transfer the sweat from body to atmosphere. In polymeric material, the water vapour transmission is performed in different steps, namely

(a) Adsorption of moisture on the surface
(b) Solubilisation within the polymer matrix
(c) Moisture diffusion into the structure
(d) Desorption of moisture on the other side of the polymer.

These steps will happen perfectly in the case of hydrophobic polymers. But in the case of hydrophilic material, this transmission mainly depends upon the amount of water activity [39]. The water vapour permeability of the bacterial cellulose in the wet state is analysed by Wen-Chun Lin et al. The results of the study revealed that the initial moisture loss from the wet cellulose is noted higher and then it is reduced. After analysis, the study reported a water vapour transmission rate of 1503 g/m^2/day in the native bacterial cellulose [38]. The moisture absorption isotherm study of bacterial cellulose reported that increment in moisture content increases the water vapour transmission through cellulose material. The initially absorbed moisture reacts with the amorphous region which results in the swelling of material. This water molecule acts as a plasticiser and opens the structure of the matrix. On further increment of moisture content, the water vapour transmission stays constant. This could be attributed to the saturation in reactive sites or amorphous region. Due to the saturation, the transmission of the permeability stabilises after certain moisture content percentage [35].

In the dry bacterial cellulose, the addition of functional material increases the permeability. The water vapour permeability was increased with bacterial cellulose–alginate composite material. The permeability of the native bacterial cellulose is noted as 0.25 g/cm^2/day, and with the addition of alginate, there was a 2.1% increment noted. A maximum water vapour permeability of 0.51 g/cm^2/day is noted for 20% alginate. The further addition did not have any impact on the permeability [40]. The dry bacterial cellulose–chitosan–polyvinyl alcohol composite film showed a moisture transmission of 1.86×10^{-12} to 1.17×10^{-11} g/m Pa. As the atmosphere or sample moisture content increases, the initial absorption and monolayer formation of the film showed a sudden increment in transmission due to the plasticising effect caused by swelling [41]. Similarly, the dried native bacterial cellulose developed from *A. xylinum* was analysed and compared with textile material. The results reported that the water vapour permeability value of 151 g/m^2/day is noted and the researcher mentioned that this is far lower than a normal cotton fabric (437 g/m^2/day). These results are supported by the reason that the porous nature of the dried bacterial cellulose material is lower than the wet one. The structural change in the drying process reduced the pore size, and so the water was not able to penetrate into the structure as reported in the wet state. However, the researcher validated their results by reporting that the permeability value of the dried bacterial cellulose is closer to the normal skin vapour transmission value of 204 g/m^2/day [10].

3.6.3 Wettability or Water Contact Angle

The contact angle value represents the surface wettability of the material. The measurement is commonly used to measure the hydrophilic and hydrophobic nature of the textile and polymeric material. The contact angle values differ from 0° to 180°, by representing complete hydrophilicity to hydrophobicity, respectively. Water and liquid contact angle of the *Gluconacetobacter xylinus*-based bacterial cellulose was evaluated after freeze-drying method. In a study, the researcher compared the native cellulose and chemically modified cellulose. The findings revealed that control native bacterial cellulose had a water contact angle of 49°, 39°, 47° and 41°, respectively, for water, formamide, ethylene glycol and diiodomethane. This represents a higher hydrophilic nature of the bacterial cellulose. However, the esterification process increased the water contact angle and showed a higher hydrophobic nature of the chemically modified cellulose. The major reason for the increment in the water contact angle is due to the lower surface energy after chemical treatment [42].

On room-dried bacterial cellulose, the water contact angle is noted as 28° on average of multiple analysis. Any contact angle less than 45° represents good wettability and 0° represents complete wetting. The researcher mentioned that the post-treatment of the native cellulose by NaOH might be the major reason for the hydrophilic nature. Next to that, the structural parameters like surface roughness and higher amorphous region are important factors [10]. Along with in-situ and ex-situ modification, the surface functionalisation methods can also be used for modification. Plasma treatment and silanisation with trichloromethyl silane on the surface of the bacterial cellulose were performed, and the effect on the water contact angle was performed. The results found that plasma treatment improved the hydrophilic nature of the bacterial cellulose by further reducing the contact angle of water. The silanisation process is a chemical vapour deposition process. Hence, the surface energy greatly reduced and the bacterial cellulose is converted into hydrophobic. Similarly, the silanisation cum plasma treatment converted the sheet more hydrophobic in both wet and dry states from an angle of 24.9° to 132.6° [43]. Table 3.2 consolidates the various physical and chemical properties of bacterial cellulose from the literature analysis.

3.7 Purification and Bleaching of Bacterial Cellulose

Bacterial cellulose produced from different sources has different colours and structural properties. One of the standard media for cellulose production from bacteria is HS medium that basically yields a dull white cellulose mat. Similarly, as discussed in the previous chapter, several alternative carbon and nitrogen sources along with various nutrients can yield different amounts of bacterial cellulose in different colours based on its components. When it comes to fashion and textile application, colour and aesthetic appearance of the material is one of the very important properties a material should possess. The main interest towards the bacterial cellulose is due to its bio-based cellulose with higher purity, sustainable nature, mechanical and other unique natures as discussed in the previous sections. Traditionally, the cellulose is always subjected to the purification process to increase its quality and also to obtain good results in the subsequent wet processing stages. In similar aspects, the bacterial cellulose is also subjected to purification for the removal of its media components and other biomatters like dead bacterial cells, etc. In this aspect, several researchers scoured the bacterial cellulose with sodium hydroxide and neutralised. Further bleaching and colouration of the bacterial cellulose are also performed to increase its commercial potential and consumer acceptance.

Other than the aesthetic look and physical characteristics, the removal of remaining biomaterial and non-cellulosic components (typically known as impurities) like living or dead organisms, very low molecular sugar components and other nutrients that help in metabolic activity of the bacteria are essential. Presence

Table 3.2 Fundamentals of bacterial cellulose properties

S. no.	Properties	Structure	Effect of modification
1.	Morphological properties	• Nanofibrils • Highly porous structure • More open structure • Higher pore volume and the surface area identified	• Ex-situ method-added particles are coating on fibre or deposits or aggregates on and in the structure • In-situ or ex-situ modifications increase the fibre diameter • Reduces the porosity. However, based on the material used, the porosity changes • The structure becomes compact and denser • Significant reduction in pore volume and surface area
2.	Chemical properties	• Bacterial cellulose consists of cellulose I and cellulose II structure • The FTIR spectrum of the bacterial cellulose is similar to the plant-based cellulose • The bacterial cellulose is higher crystalline than any other type of cellulose • The typical XRD diffraction peaks of bacterial cellulose noted at 14, 16 and 22° wavelength	• The type of source used and the additives used in the cultivation process decides the structure and chemical properties • Chemical treatments like alkali and bleach increase crystallinity • Ex-situ modification alters the crystallinity based on the type of chemicals used
3.	Thermal properties	• The ideal bacterial cellulose shows three differential sections in thermal degradation • Initial moisture loss, loss of low molecular weight components and melting of cellulose at the final stage • High glass transition temperature and the melting temperature were noted • Noted stable for higher temperature and thermal degradation above 350 °C	• Chemical modifications like alkali treatment reduced the low molecular weight components in the cellulose structure. Hence, the modified cellulose showed a two-step degradation • The chemically treated bacterial cellulose showed a higher temperature withstanding capacity due to the removal of impurities • The spectrum showed a broad peak and longer temperature stability • In-situ and ex-situ modification also affected the temperature ranges based on chemicals treated

(continued)

Table 3.2 (continued)

S. no.	Properties	Structure	Effect of modification
4.	Physical properties	• Due to higher crystallinity and degree of polymerisation, the bacterial cellulose has higher tensile strength and Young's modulus • Lower extension percentage noted in the native cellulose	• The chemical modification process/purification process generally increases the crystallinity and so the mechanical properties • In-situ and ex-situ modifications either increase or reduce the properties based on the type of material used in the modification
5.	Moisture-related properties	• Free and open structure offers more free sites to engage free water molecule and provide higher water absorption • 200–400% of water holding capacity with a more or less equal amount of retention capacity at both dry and wet states • Highly rough surface area and hydrophilic surface with very lesser water contact angle at wet and dry state • Water vapour permeability is noted high at the wet state and lower at dry state	• In-situ modifications had reduced the water holding capacity, vapour transmission capacity due to the increment in fibre diameter and structural compactness • However, highly hydrophilic functional material showed an increment • Ex-situ modification did not have any significant effect on the moisture properties • Based on the chemical used, functionalisation can be done. For example, the cellulose mat can be converted into hydrophobic after treatment

of these components on the structure may potentially lead to the contamination of fabric with other micro-organisms later, and that may create unpleasant odour from the fabric. Based on these points, it can be understood that purification is one of the essential requirements for fashion and textile application. In this section, different post-treatments performed on bacterial cellulose and its effect on properties were detailed exhaustively.

3.7.1 *Purification of Bacterial Cellulose*

Sodium hydroxide (NaOH) or alkali is the common material used in the purification of plant cellulose. Alkali mainly penetrates the structure in the amorphous region of the cellulose, hydrolyses the impurities in the cellulose and helps in forming new crystalline lattices after wash-off. The general procedure for the NaOH treatment of bacterial cellulose is listed in Fig. 3.12. Once the cellulose matured in the culture medium after the fermentation time, the NaOH treatment helps in removing non-cellulosic items including the proteins and nucleic acids present in it. To convert the

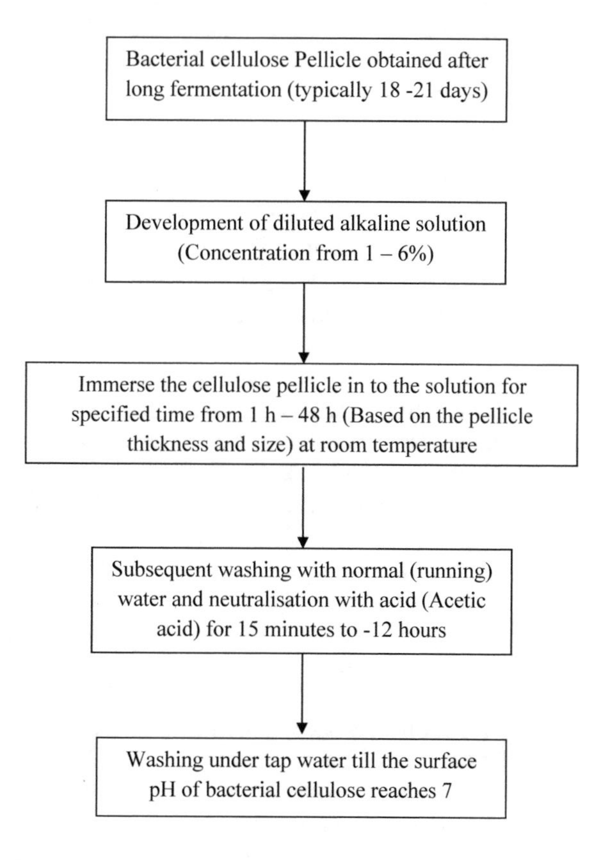

Fig. 3.12 Process steps involved in the alkali treatment of bacterial cellulose [Authors own illustration]

bacterial cellulose sheet into a useful fabric by successive post-treatment, Juyeon Han et al. treated the bacterial cellulose pellicle with various concentrations of the NaOH from 0 to 4%. Their results reported that the higher the concentration of the alkali better will be the removal of impurities from the cellulose mat. They have reported the removal of dead bacterial cells from the mat using SEM micrographs [18]. At the same time, the researcher also mentioned that the use of higher concentration alkali will shrink the cellulose mat and cause curling in the cellulose mat. This is due to the transition of cellulose crystal structure from cellulose I to cellulose II. Gae et al. reported that the changes in the cellulose are mainly due to the alkalisation process. This process breaks inter- and intramolecular primary hydrogen bonds that are naturally present in the cellulose [32].

Researchers also analysed the effect of NaOH concentration on the shrinkage percentage. Upon increment from 2% NaOH, the result showed a drastic shrinkage up to 4% and further increment showed a reduction in the shrinkage. However, at 10% concentration maximum shrinkage was noted with the bacterial cellulose. The researcher also compared the effect of potassium hydroxide (KOH) on the shrinkage properties. The results revealed that the shrinkage levels were less compared to the NaOH [44]. In a detailed study on the effect of NaOH treatment on cellulose and its structural analysis, researchers evaluated the effect of NaOH concentration from 0 to 30% concentration and measured the cellulose II formation percentage. The results revealed that from concentration percentage 10–16 there was a sharp increment in the cellulose II formation. At 16% NaOH, around 75% of the cellulose in the structure is converted into cellulose II structure. They have also reported the at 25% and above, the cellulose formation is noted in the first 2 min of the reaction and on further there is no cellulose II formation noted due to the structural degradation [45]. Hence, several researchers used the NaOH concentration less than 10% most of the time to achieve better cleaning with minimal cellulose II formation in the structure to retain structural integrity. The detailed mechanism was discussed by Okano and Sarko. They reported that at the first stage the treatment induces Na-cellulose I in the internal structure of the cellulose. In the second stage, the antiparallel structure of cellulose II is produced from the Na-cellulose I. Once the alkali enters into the cellulose structure's amorphous regions, it converts the amorphous sections into antiparallel Na-cellulose I. On continuous treatment, the alkali converts the crystalline region also into antiparallel Na-cellulose I with alkali interface. On subsequent treatment, the overall amorphous region increases and forms Na-cellulose II in the structure. Upon washing and drying, the NaOH is completely removed and the cellulose II forms as a result of the alkalisation process [46]. The illustrative possible mechanism of the alkalisation is provided in Fig. 3.13.

One of the most important parameters next to the alkali concentration is the treatment time. Several researchers noted the effect of NaOH treatment on the various times from 5 min to 48 h as reported by Gea [47]. He reported that at the same concentration, the treatment of bacterial cellulose for 24 h had the same effect of the treatment with a shorter time of 2 h. In order to avoid the formation of cellulose II, by alkalisation, Gea developed a two-stage treatment process of bacterial cellulose that provides a 2.5% NaOH treatment and consecutive 2.5% NaOCl overnight. The researcher

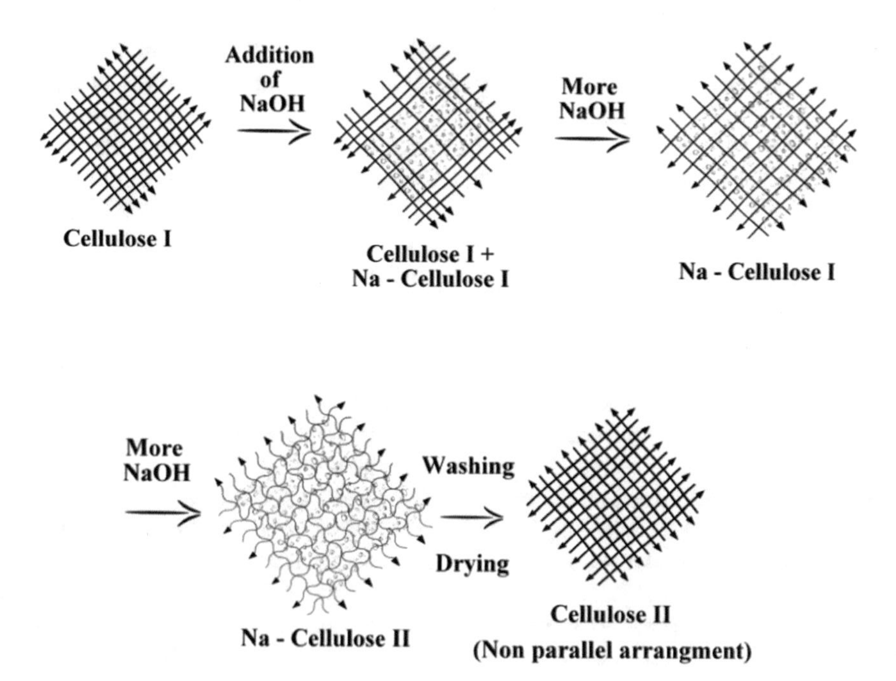

Fig. 3.13 Mechanism represents the changes in bacterial cellulose under alkali treatment [46] [Authors own illustration]

compared the two-step purifications along with NaOH-treated and NaOH-untreated bacterial cellulose. The findings revealed that the two-step process increased Young's modulus and tensile strength value of the bacterial cellulose than the single-step and untreated bacterial cellulose. The electron micrograph images also confirmed the improved removal of the non-cellulosic content from the three-dimensional structure [47].

(i) **Effect on Mechanical Properties**

Irrespective of the application process, the removal of biomaterial from the produced cellulose is an important step after the production of bacterial cellulose. Hence, several research workers attempted different methods and chemicals for the same. One of the important drawbacks with the purification process is higher concentration, and longer time will degrade the cellulose structure and result in a loss in structural integrity. Kamal et al. analysed the effect of NaOH percentage on the properties of the bacterial cellulose produced from *A. xylinum* using coconut water as a medium. The produced bacterial cellulose is treated with different concentrations, namely 0, 0.5, 1 and 2%. The chemical analysis by FTIR spectrum reported the presence of cellulose I structure even after the 2% NaOH treatment. However, the crystallinity percentage of the 1% NaOH-treated bacterial cellulose is noted higher (79%) than that of other concentrations. On the morphological analysis, higher concentration of alkali showed effective removal of impurities and also at 2% concentration, fibre swelling was reported by the author [48]. Xinyu Wu detailed the effect of alkali concentration on the mechanical properties of the bacterial cellulose. The researcher noted a reduction in the strength of the cellulose matrix with an increase in the NaOH concentration. Out of all the cellulose mat, untreated bacterial cellulose mat had higher strength. Out of alkali treatment, 1% concentration of NaOH-treated bacterial cellulose showed a maximum strength, which is approximately 64% of the untreated cellulose mat. The reduction in tensile strength gradually increased the elongation and toughness of the NaOH-treated bacterial cellulose mat from 0 to 7% concentrations [44].

The main reason for the changes in the mechanical properties of the bacterial cellulose is due to the transition of the cellulose structure and due to the alkalisation process. The conversion of cellulose I to II increases the flexibility and toughness; however, the tensile values reduce due to the molecular changes. The structure of the cellulose initially consists of covalent bond for parallel aggregation and is cross-linked by a hydrogen bond. The alkali treatment technically disturbs inter- and intramolecular bonds. After the removal of NaOH by washing, the cellulose arrangement disorients and forms thermally stable cellulose II structure. The removal of NaOH provides more spaces between the molecules, and it leads to increased mobility of the molecular chains. This enables better mobility to structure and provides increased flexibility [44].

(ii) **Porosity and Transmission Properties**

Tito Arif Sutrisno et al. analysed the pineapple peel extract-based bacterial cellulose for its permeability properties before and after the various degrees of alkali treatment. The researcher treated the 14 days fermented cellulose with 2.5, 5 and 7.5% alkali. The results of the study revealed that the alkali treatment increased the interfibre space and reduced the fibre size. The surface roughness increased due to the formation of white crystal on the surface by the deposition of alkali. The porosity of the untreated cellulose is noted as 19.5%. The alkali treatment increased the porosity to 50.94%, 61.90% and 87.13%, respectively, for the 2.5%, 5% and 7.5% NaOH treatment [49]. Other researcher analysed the effect of NaOH treatment on water vapour permeability of bacterial cellulose membrane. Upon increase in NaOH concentration, water vapour permeability of the bacterial cellulose matrix increases due to the formation of pores in the structure. The researcher noted a 21.59% increment in water vapour transmission in 0.5% NaOH-treated material and 36.5% with 1% NaOH-treated cellulose. But on further increment, at 2% NaOH, the transmission of water vapour reduced 14.14% than the 1% NaOH-treated sample. The researcher explained the reason might be due to the removal of impurities in the initial stage (0.5–1% NaOH) of concentration so the increment in pore size resulted in an improved water vapour transmission. But further addition of NaOH (2%) resulted in a fibre swelling and so a reduction in the interfibre space. Thus, a higher concentration of NaOH reduced the water vapour transmission significantly [48]. After the treatment of 0.1% NaOH, Meftahi et al. reported that the NaOH treatment has increased the water droplet absorbency time from 230 to 270 s in the dry state. Similarly, the researcher also noted a 10% reduction in water absorption capacity of the purified cellulose after the NaOH treatment. The removal of media components and other impurities are noted as the main reason for the reduction in water absorption and holding capacity of the bacterial cellulose [50]. Elham Esmaeel Al-Shamary et al. analysed the porosity of the bacterial cellulose after treating with different alkalis, namely sodium hydroxide, potassium hydroxide, sodium carbonate or potassium carbonate. Their results reported that NaOH has produced a very less porosity of 64% than all the alkalis used. A maximum porosity of 75% is obtained in the case of potassium carbonate. The researchers reported that the higher swelling nature of the NaOH must be the reason behind the lower porous nature of the NaOH treated bacterial celluose compared to other alkali treated samples [51].

(iii) **Effect on Thermal Properties**

Alkali treatment of bacterial cellulose developed from coconut water and *A. xylinum* was performed by Lucia Indrarti and Rike Yudianti. They evaluated the effect of alkali on the thermal behaviour using thermo gravimetric analysis and reported that the alkali treatment completely changed the thermal degradation pattern of the bacterial cellulose. The degradation percentage in the second stage due to thermal degradation is 67.1% in the case of alkali-treated and 65% in the case of non-treated [52]. The thermal degradation analysis of untreated bacterial cellulose, NaOH-treated and NaOH and NaOCl (two-step method)-treated was done. The results revealed that the degradation temperature of the NaOH-treated bacterial cellulose increased to 322 °C from 289 °C of untreated. In the case of two-step treated, the degradation temperature is noted as 359 °C. Similarly, the amount of residual mass % at 1000 °C is reduced when the purification process increases. For untreated, the mass % is noted as 33.97%, NaOH-treated is noted as 28.11% and two-step treated showed a mass % of 14.54%. The lower degradation temperature and higher residual mass associated with the untreated cellulose are mainly associated with the presence of impurities. After purification, the residual mass reduced significantly and degradation temperature increment showed significant thermal stability of the treated bacterial cellulose [47].

Other researchers analysed the effect of various alkalis on the thermal characteristics of the bacterial cellulose produced from *A. xylinum* in a sucrose medium. They evaluated the effect of NaOH, KOH, Na_2CO_3 and K_2CO_3 by treating it 30 min at boiling temperature. Differential scanning calorimetry results reported that the crystalline melting endothermic peak of native cellulose is noted in the range of 120.47 °C. However, after treatment, there is a shift in the temperature of the crystalline melting process of bacterial cellulose mat to 106.48 °C with NaOH. Concerning KOH, the temperature is noted as 109.04 °C, and for Na_2CO_3 and K_2CO_3 it is noted as 109.65 °C and 112.61 °C. This shift is associated with the changes in the crystalline nature of the mat and the conversion of cellulose I to II. Similarly, at higher temperature, the degradation of the native cellulose mat is noted as 294.07 °C. But in the case of alkali-treated material, the stability increased and the temperature is noted as 343.27 °C for NaOH, 352.67 °C for KOH, 357.45 °C for Na_2CO_3 and 370.06 °C for K_2CO_3. To evaluate the thermal degradation behaviour better, the thermo-gravimetric analysis was performed. In the case of untreated native cellulose, around 15–20% of the weight loss in the temperature range of 250–285 °C is noted. For other types of alkali-treated samples, the degradation order was noted as $K_2CO_3 > Na_2CO_3 >$ KOH > NaOH between a temperature range of 328.75 °C and 337.63 °C. The increase in decomposition temperature suggests that the alkali treatment removed the low molecular weight impurities and developed changes in the crystalline nature of the cellulose material [25].

3.7.2 Bleaching of Bacterial Cellulose

The general steps involved in the bleaching process are illustrated in Fig. 3.14. Next to the NaOH treatment, one of the highly adapted chemical treatments is the bleaching process. The process is performed to remove the unwanted colouring matter present in it. The main aim of the bleaching process is to enhance the whiteness index of the cellulose pellicle. The whiter the cellulose pellicle, better will be its application potential in the fashion and textile industries. This is because the bleached cellulose can also be used for further colouration in the wet or in the dry state to enhance its aesthetic properties. Hydrogen peroxide is one of the most commonly used bleaching agents. Juyeon Han et al. analysed the effect of different concentrations of hydrogen peroxide (1–30%) on bleaching efficiency. With the different

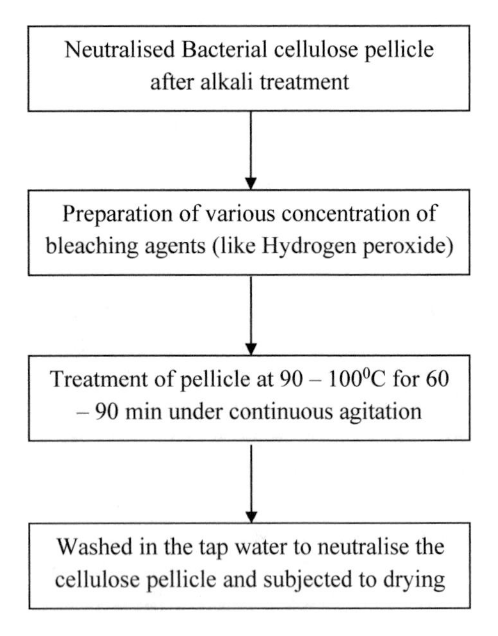

Fig. 3.14 Steps involved in the bleaching process of bacterial cellulose [Authors own illustration]

concentrations of hydrogen peroxide, a maximum whiteness index of 73.15% is noted for 5% concentration. In the case of higher concentration like 10 and 20%, no much difference is noted with the whiteness index. But there is a reduction in 30% concentration noted. The researcher reported that at a higher concentration above 5% the structural degradation was noted and that is the main reason for changes in the whiteness index. Furthermore, a higher concentration of bleach also damages the cellulose fibre structure and that was proved by the electron micrographs. They selected 5% peroxide as the optimal concentration for bacterial cellulose bleaching without structural damage. The crystallinity analysis report revealed that there was a significant amount of crystalline percentage increment in the bleached matrix. Crystallinity percentage of 53.3% increased to 73% after alkali treatment and further increased to 80.1% after bleaching. The researcher reported the bleached cellulose can be further dyed and finished for textile applications [18]. As like the alkali treatment, the bleaching process also has a significant impact on the mechanical, thermal and structural properties of the bacterial cellulose.

Researcher treated the developed bacterial cellulose with sodium chloride instead of hydrogen peroxide. The study evaluated the impact of bleaching process on mechanical, chemical and thermal properties of the bacterial cellulose. The results reported the treatment significantly increases Young's modulus of the cellulose mat than the alkali treatment. The similar improvement is also noted in the tensile strength value from 139.1 to 207 Mpa. The improvement suggested that the effective crystallisation of the cellulose mat and the removal of impurities better than alkali treatment are the main reasons. Upon thermal degradation analysis, the degradation temperature increased further than the alkali-treated cellulose. In alkali-treated bacterial cellulose, the degradation temperature is noted as 322 °C, and in the case of bleached cellulose, it is noted as 359 °C. The chemical analysis on crystalline structure revealed that on the bleaching process, no conversion of cellulose II happened; instead, more alignment or orientation of fibrils increases the crystalline nature of the bleached cellulose [47]. HeruSuryanto et al. analysed the changes in crystallinity percentage of bacterial cellulose with different concentrations of hydrogen peroxide. The crystallinity percentage of the cellulose increased with the increment in the NaOH percentage from 0 to 5%. However, there is a reduction noted in 7.5%. The effective removal of amorphous regions from the structure is the main reason for higher crystalline nature at 5% (85.1%) NaOH than untreated one (83%) [53]. Firdaus Jahan isolated the *A. xylinum* from rotten fruits and analysed its properties. Among the selected bleaching agents, namely calcium hypochlorite is found to be more effective on the bleaching process than the sodium hypochlorite and H_2O_2. Hence, the researcher optimised the concentration and bleaching time. The results reported that at 2% concentration the cellulose bleached 90–95% and further increment in the calcium hypochlorite concentration reduced the strength of the material. Similarly, to achieve a better whiteness index 8–10 h is noted as the optimum time for calcium hypochlorite bleach [54]. Other researchers evaluated the effect of alkali and bleaching process individually on the tensile and porosity of the bacterial cellulose. Though a significant reduction in tensile strength noted with alkali treatment (33.02 Mpa), there is an increment in tensile strength noted with peroxide treatment (44.62 Mpa). With

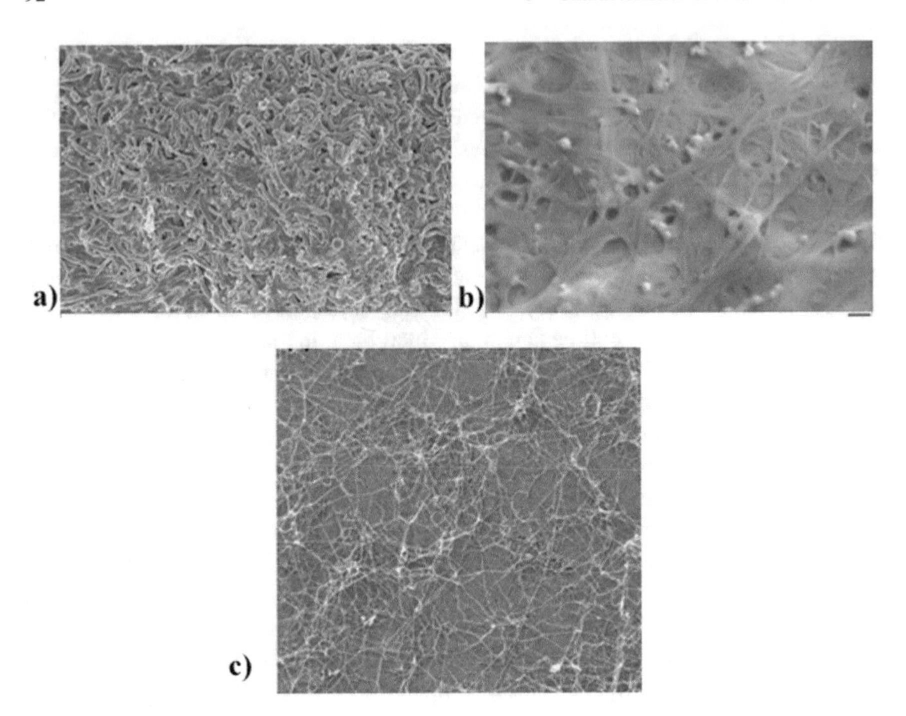

Fig. 3.15 Scanning electron microscope view of **a** raw bacterial cellulose (with impurities and dead bacteria) and **b** alkali-treated [Authors own illustration] and **c** bleached [Reprinted with Permission]

respect to the porosity, the findings revealed that alkali treatment has produced more pores in bacterial cellulose than the peroxide treatment [49]. Figure 3.15 represents the structural changes in the bacterial cellulose mat after the alkali and bleaching process. In Fig. 3.15 (a), the untreated bacterial cellulose with dead bacteria and media components can be noted. In Fig. 3.15 (b), the NaOH-treated cellulose shows a clear image of the fibril structure; however, there are still many other components present on the surface, and (c) shows the bleached cellulose structure where except the cellulose fibre all the impurities were removed.

In the textile view point, few researchers produced the bacterial cellulose and analysed its properties. Gayathry and Gopalaswamy developed a bacterial cellulose mat using HS medium and *A. xylinum.* The developed cellulose mat was treated with 0.5% NaOH for 30 min and further bleached with 1.5% hydrogen peroxide. The bleached fabric was compressed and allowed to dry, and then the suitability towards the textile application was evaluated and reported. They analysed tensile strength, moisture content and water holding capacity and reported the suitability of bleached bacterial cellulose in textile application [55]. Similarly, other researchers analysed the use of *A. xylinum*-based bacterial cellulose for home textile application. The researcher bleached the bacterial cellulose with 10% bleach after 24 h immersion in 1% NaOH. The researcher developed an ottoman and analysed its sewability of the developed bacterial cellulose. The results were positive, and wrinkle appearance of the bleached cellulose is noted as the only disadvantage with the developed product [56]. Though several research workers analysed the chemical and physical changes in the bacterial cellulose structure through direct bleaching or by subsequent bleach after NaOH treatment, the researches were preliminary and apart from the fundamental research, the work on bleached material in the textile and fashion field is still meagre. Very few researchers performed so far in the textile and fashion applications [18, 55].

3.8 Colouration Properties

Dyeing or colouration is one of the necessary steps in the fashion and textile product manufacturing. As fashion products are highly trend-oriented, the customer requirements also changes in a faster manner concerning time. As the bacterial cellulose is proposed for textile and fashion applications, the possibilities of colouration also become one of the important requirements. As the application of the bacterial cellulose in fashion is very new and initiated a couple of years ago, the amount of research performed on the colouration of bacterial cellulose is meagre. However, in recent years, after 2015 there are few research works performed around the world to increase the aesthetic property. Bacterial cellulose commonly uses HS medium as a source, which yields cellulose in the yellow colour due to the presence of agar and other media components.

Other than HS medium, few other common fermentation media offer its colour to the produced bacterial cellulose. For example, bacterial cellulose production with the help of 'tea' using kombucha-based SCOBY is widely adapted in the apparel and fashion field due to its easiness in preparation. The method preferably offers the tone of the tannin present in the tea. The first research on coloured bacterial cellulose in apparel end use was performed by designer Suzanne Lee. She is one of the leading fashion designers who works primarily on the bio-based alternative raw material for the fashion sector. Based on her research, she founded a consultancy named 'BioCouture' to produce fabrics from bacterial cellulose. She developed bacterial cellulose and used it for apparel and accessory development without any colouration with the same tea colour. The product is named as bioleather, and it gained a lot of attentions among the environmentalists and received a great appreciation by mentioning this could be an alternative to animal leather [57]. The garment designed by Suzanne Lee using kombucha tea-based bacterial cellulose fabric is provided in Fig. 3.16b [57].

Few other research workers also tried to produce coloured bacterial cellulose with tea and reported the coloured cellulose as vegan leather used it as an alternative material for leather accessories like shoe and handbag. Armine Ghalachyan developed bacterial cellulose from the tea, sugar and vinegar along with the commercially available kombucha SCOBY culture. The kombucha based bacterial cellulose used for the development of bacterial cellulose-based bags and shoes as a leather alternative. The half-tone tannin look and dried surface wrinkle like structures made the bacterial cellulose to look like a leather material [58]. But when it comes to regular apparel, the consumer preferences are different and vary based on several factors.

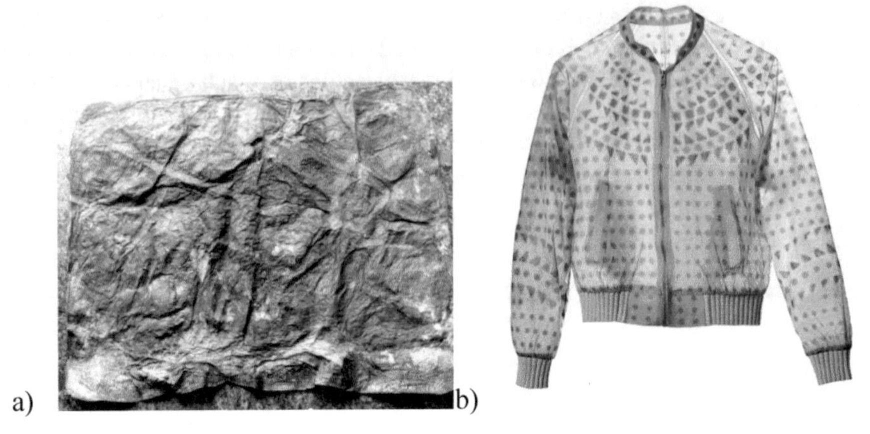

Fig. 3.16 **a** Raw bacterial cellulose from tea and kombucha. **b** Jacket prepared using kombucha-based bacterial cellulose by designer Suzanne Lee [57]

Hence, the brown or tannin colour produced by the tea is very insufficient to fulfil the fashion trend requirements. Recently, researchers analysed the various mediums like red wine, beer, milk, black tea, green tea and coconut juice to develop bacterial cellulose. The primary aim of the research was mainly focused on the analysis of moulding ability of the developed cellulose on the body form. However, we can also see from their output that the different sources also produced bacterial cellulose in a different colour based on its native colour. For instance, the red wine and black tea yielded a brown colour, beer yielded a yellow, and milk and coconut juice produced white bacterial cellulose [59]. Figure 3.17 represents the schematic pathway of (a) in-situ and (b) ex-situ coloured bacterial cellulose production.

3.8.1 In-Situ Dyeing of Bacterial Cellulose

Though it was one of the default processes, obtaining the desired colour is difficult. Zhong Chunyan patented the process of colouring the bacterial cellulose in situ with additional colourants. In his patent, the food colour is introduced inside the inoculums or bacterial cellulose production medium. The researcher also included chitosan along with the normal media component and patented the process. The addition of chitosan will help in reducing the negative charges of the fibre and avoid the repulsion of pigments. The researcher proposed different recipes for different colours in the patent, and it is the first of its kind known as in-situ colouration [60]. Tyurin et al. prepared the bacterial cellulose using *A. xylinum* and osmophilic yeast strains for a different period of 1 and 2 week time. They included American blueberry extract into the culture medium as a natural dye and cultivated the cellulose. The results revealed that the blueberry extract induced a ruby-red colour to the developed

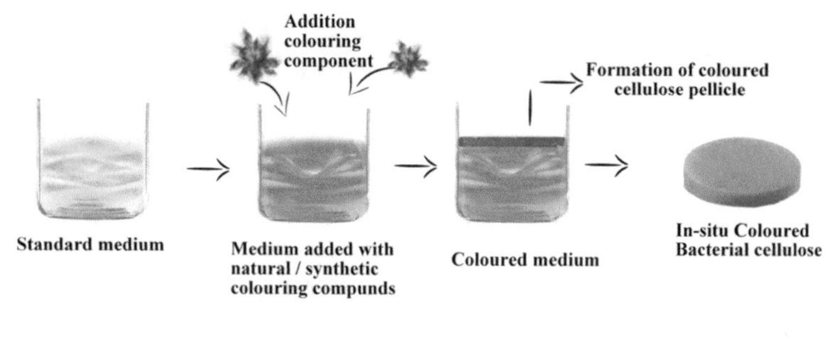

a)

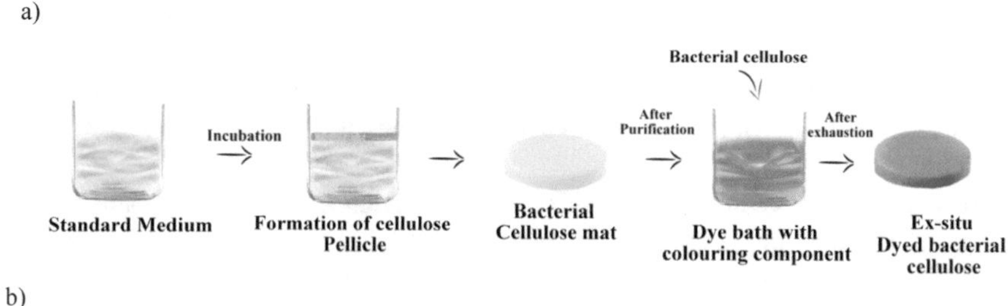

b)

Fig. 3.17 Schematic illustration of **a** in-situ and **b** ex-situ colouration of bacterial cellulose [Authors own illustration]

bacterial cellulose. After removal, the researcher analysed the mechanical properties after a cleaning process with 2% NaOH. The results revealed that a higher strength of 7.3 Mpa noted for 2 weeks fermented cellulose over a week fermented (5.2 Mpa) cellulose. Similarly, the elongation of the 2-week fermented bacterial cellulose is noted less (14.9%) than the 1-week fermented cellulose (19.3%) [61] (Fig. 3.18).

Lv et al. reported a concept paper by in-situ dyeing of bacterial cellulose using natural dyes. They proposed potential possibilities of dyeing of bacterial cellulose

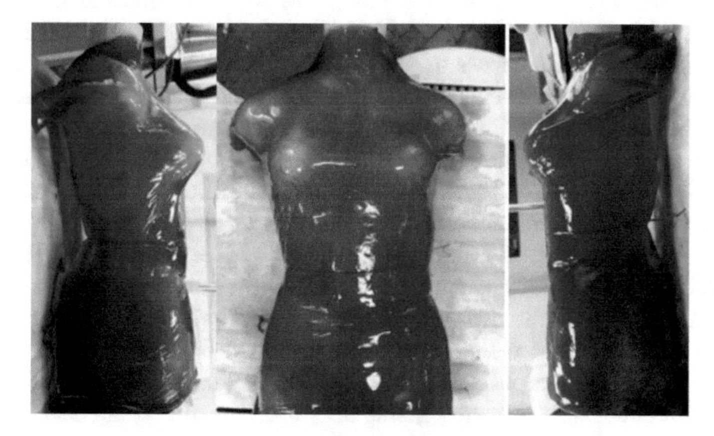

Fig. 3.18 In-situ dyed bacterial cellulose with blueberry extract [Reprinted with Permission]

and usage of cellulose film in the packing film [62]. In the phase of commercialising the bacterial cellulose non-woven for the textile applications, other researcher dyed the bacterial cellulose in situ using natural/synthetic colouring substances like turmeric, saffron and beet and artificial colours including red, orange, yellow, green, blue and voile. After a three-week fermentation, they analysed the thickness (mm), tensile load (N), elongation (%), tearing strength (N) and stiffness of the developed coloured bacterial cellulose. Based on the results, they had concluded that the developed coloured cellulose non-woven material had a higher amount of the tensile strength (199.2 N) and elongation (27.1%). Stiffness (133.7 mg) and tearing strength (5.49 N) are noted comparable with the commercial non-woven existing in the market. However, the fabric was bit heavier (weight of 840.8 g/m^2) than the commercial fabric due to its higher water content [63]. However, they did not report any detailed analysis of the individual dyes and among the natural and synthetic dyed bacterial cellulose.

Researchers analysed the possibility of dyeing bacterial cellulose in-situ and ex-situ method using direct, acid and reactive dyes. The selection of dyes was based on their commercial applications and also based on different interaction mechanisms, namely aggregation, ionic bonds and covalent bonds. SCOBY-based bacterial cellulose is produced using HS medium after 8 days fermentation, washed with 3% NaOH and neutralised with acetic acid. In the case of in-situ dyeing, 0.015% dyestuff is added inside the culture medium. After fermentation, the researchers noted that produced bacterial cellulose in the reactive dyed medium is twice the amount of native cellulose medium. The main issue noted by the researcher is the produced cellulose coloured only at the bottom of the pellicle, and it is noted that there is no colour at the top. This is mainly due to the aerobic nature of cellulose production. However, they reported the same colour on both sides upon drying. The colour difference in the top and bottom side of the bacterial cellulose is mainly associated with the different densities of the cellulose produced in the top of the layer and bottom. Higher density in the upper side restricts the penetration of dyestuff to the top side. In order to confirm the same, the crystallinity analysis performed and noted that the top side of the cellulose pellicle is highly crystalline (73.6%) than the bottom side (68.5%). The surface roughness of the in-situ dyed cellulose reduced to a greater extent from 880 to 194 nm due to the filling of surface cavities in the microfibrils with dye particles [64].

In comparison with in-situ and ex-situ dyeing method, it is reported that the higher colour strength is noted in the case of in situ (double time) than ex situ. The in-situ method offers higher surface smoothness and reduced manufacturing sequence and time. Due to the additional potentiality of the in-situ method including surface character alteration and reduction in the time, the researcher proposed in situ is the better option for the dyeability of the bacterial cellulose [64]. In-situ functionalisation of bacterial cellulose is also performed by other researchers. Few researchers analysed the possibility of modifying or functionalising the bacterial cellulose with fluorescence characteristic. In this process, the researcher used modified glucose with 6-carboxyfluorescein (6CF) as a feedstock material in the production using *Komagataeibacter sucrofermentans*. Upon fermentation, the cellulose is analysed with the chemically modified cellulose and found that the biologically modified cellulose had a green fluorescence looking under UV lamp. In comparison with the chemically modified cellulose, the developed cellulose had superior fluorescence intensity. A further chemical analysis reported that no chemical changes in the basic structure of the modified bacterial cellulose. The researcher reported that based on the concentration used, the properties of the developed cellulose also can be controlled. As compared to the chemical usage and modifying method, the 6-carboxyfluorescein (6CF) included bacterial cellulose found to be environment-friendly and cost-effective [65].

3.8.2 Ex-Situ Dyeing of Bacterial Cellulose

The second method of colouring the bacterial cellulose is ex-situ dyeing. In this method, the bacterial cellulose is initially produced and then dyed using a colouring medium in the later post-processing stage. Harmon produced bacterial cellulose and dyed it with beet extract and vinegar. The researcher developed two different types of colour and used it for the upper and lower part of the body [66]. The cellulose developed using glucose and corn steep liquor medium using *G. hansenii* was dyed with *Clitoria ternatea* L. and *Hibiscus rosa-sinensis* flower extract. The dye extract was treated with cellulose pellicle with sodium chloride and sodium ferrocyanide as a mordant for 30 min at boiling temperature. After dyeing, cellulose pellicle is treated with 1% dye-fixing agent and 2% softener and neutralised for the pH of 7. After that, the cellulose pellicle is washed in tap water and immersed in the deionised water till the removal of excess colours on the surface. The results were compared with the ARAQCEL RL 500, a commercial indigo blue dye. The findings revealed that the penetration of the commercial dye was noted good than the natural dye. The chemical structure analysis revealed an 84.37% crystallinity of the native nacellulose. However, a reduction in crystallinity is noted with the dyed sample 82.23 ± 0.21, 80.72 ± 0.19 and $73.22 \pm 0.15\%$, respectively, for *C. ternatea* L., *H. rosa-sinensis* and ARAQCEL RL 500. This is mainly attributed to the reduction in intermolecular force due to the addition of dye molecules inside the structure of the bacterial cellulose. Though no big difference was noted in the thermal degradation temperature, there are various amounts (increased) of weight loss noted for dyed samples than the native sample. A similar trend is noted in the case of tensile properties and Young's modulus. There is a great amount of reduction in tensile strength of the dyed sample and also breaking elongation. The findings revealed that compared to synthetic dye natural dyed material has higher flexibility in the bacterial cellulose [67].

The possibility of dyeing the bacterial cellulose through indirect or passive method is also evaluated by researchers. In this study, the researchers developed bacterial cellulose using ATCC *A. xylinus* strain in a modified molasses–mannitol media. After a 21-day fermentation, the developed cellulose is purified with NaOH treatment. The purified cellulose is then placed on coloured fabric and papers and incubated for 24 h

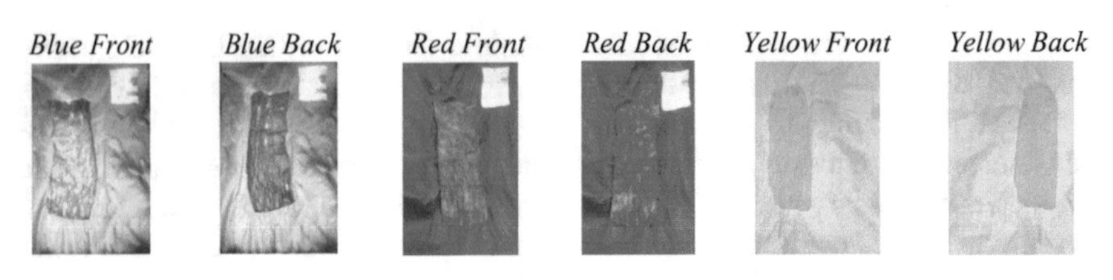

Fig. 3.19 Passive dyed bacterial cellulose material [68] [Reprinted with Permission]

to analyse potential colour transfer. The results were promising, and effective colour transfer in the case of colour paper is noted as provided in Fig. 3.19.

The results of this study reported that the dyed materials showed higher colour fastness value as overall. With respect to colour fastness to crock metre, an average rating of 4.25 is noted with a 5-point rating for red and yellow colour. In the case of wet rubbing, a colour fastness rating of 3.25, 3.75 and 4.25 is noted, respectively, for blue, red and the yellow samples. For light fastness, after 1 week of light exposure to xenon arc light, a rating of 1.75, 3.75 and 3.25 was noted for blue, red and yellow samples. The researcher pointed out the usefulness of the research for environment-friendly, waterless dyeing method [68].

Several research works are performed around the world to dye the textile with less pollution or by environment-friendly materials. In that aspect, the laccase enzyme is used to polymerise phenolic components and coat it on the conventional textile fibre surface by different methods. Based on the available information, Ji Eun Song et al. reported a biocolouration of bacterial cellulose material using laccase enzyme. The researcher developed bacterial cellulose using commercial SCOBY and using HS medium. The developed bacterial cellulose is purified with NaOH and bleached with hydrogen peroxide. After the bleaching process, the researcher immobilised the laccase enzyme into the bacterial cellulose after swelling with 8% NaOH treatment. After the inclusion of laccase in bacterial cellulose, the cellulose was cut into small pieces and immersed in selected phenolic compounds like catechol, catechin hydrate, hydroquinone, ferulic acid and quercetin dehydrate. The polymerisation process imparts the colour of phenolic compounds on cellulose fibrils. The researchers concluded that by varying initial laccase monomer concentration, the colour depth can be controlled effectively. They have developed yellow, orange, light and dark brown colour as reported in Fig. 3.20 [69].

The researcher reported that catechol, catechin, ferulic acid and hydroquinone compounds produced deep shade by laccase action. The researcher also analysed the washing durability of the coloured bacterial cellulose and noted good durability. The detailed mechanism of the colouration is unknown still but the researcher reported that the higher porosity and lower solubility of the oligomers produced during the polymerisation might be the important reason for higher durability of dyed samples [69]. In another study, an ex-situ synthetic dyeing process is performed after fermentation of the cellulose pellicle. The pellicle was treated with 0.5% dye on the weight of material with a liquor ratio of 20:1 and oven-dried at 35 °C. The cellulose is developed from different sugar mediums, namely glucose, sucrose, mannitol and fructose.

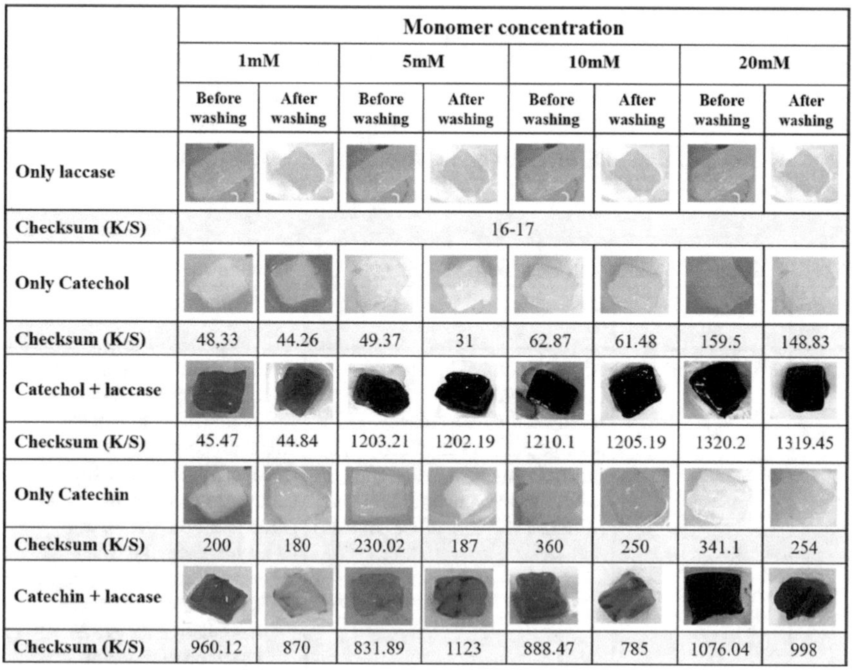

	Monomer concentration							
	1mM		5mM		10mM		20mM	
	Before washing	After washing	Before washing	After washing	Before washing	After washing	Before washing	After washing
Only laccase								
Checksum (K/S)	16-17							
Only Catechol								
Checksum (K/S)	48,33	44.26	49.37	31	62.87	61.48	159.5	148.83
Catechol + laccase								
Checksum (K/S)	45.47	44.84	1203.21	1202.19	1210.1	1205.19	1320.2	1319.45
Only Catechin								
Checksum (K/S)	200	180	230.02	187	360	250	341.1	254
Catechin + laccase								
Checksum (K/S)	960.12	870	831.89	1123	888.47	785	1076.04	998

Fig. 3.20 Bacterial cellulose samples before and after in-situ polymerisation of catechol and catechin with immobilised laccase at 50 °C, overnight; spectra values are presented as the sum of K/S values [69] [Reprinted with Permission]

On analysis, the findings revealed that the dye exhaustion percentage is higher in the case of fructose-based bacterial cellulose. The order is noted as in the sequence of fructose > mannitol > glucose > sucrose. Out of the selected dyes, the better dyeing observed for reactive dye, acid dye and least dyeability is noted for direct dye [64].

The poor dyeability of the direct dye might be attributed to the aggregation of dye molecule due to ionic charge proportion change in the dye and fibre interface. The researcher also analysed the effect of temperature and pH on dyeing ability. The results reported that the pH value of 3 (acidic) was able to create a better shade than any other pH. Similarly, a higher temperature of 135 °C yields better dye exhaustion than the lesser temperature. Researchers also reported a 10% increment in dye exhaustion on the swollen cellulose treated with NaOH. The surface analysis results by atomic force microscope revealed a significant roughness reduction in ex-situ cellulose mat (355 nm) than the native one (880 nm). In comparison with the in-situ dyed cellulose, the roughness was a bit higher [64]. Hitomi Miyamoto et al. evaluated the structural changes in the bacterial cellulose after dyeing. In this study, the researcher used *Enterobacter* species to produce bacterial cellulose and analysed its colouring ability using direct, acid and basic dyestuffs. From the results, it is noted that direct and basic dyes produced deeper shades and the acid dye did not work well with the cellulose pellicles developed. Similar to the *Acetobacter* species, there are a great number of crystallinity changes noted in the direct dyed bacterial cellulose. In the case of basic dye, no difference in the crystallinity is noted. The researcher reported that irrespective of the species types used, the direct dye reduces the crystallinity of the cellulose during the dyeing process [70]. Table 3.3 consolidates the effects of purification, bleaching and dyeing process on the bacterial cellulose structure.

Table 3.3 Effect of purification, bleaching and dyeing on bacterial cellulose

S. no.	Treatment	Mechanism	Resultant effect
1.	Alkali treatment	Alkali treatment breaks the intermolecular bonds and reorients the molecular arrangement or directions. This process converts the (cellulose I) crystalline structures into cellulose II (more of amorphous)	(i) Reduction in tensile strength and Young's modulus (ii) Increment in toughness, elongation, thermal stability (iii) Increment in porosity enhances the water vapour, air permeability of the matrix (iv) It increases the water absorbency and holding capacity (v) Resistance to thermal degradation up to 370 °C based on the type of treatment
2.	Bleaching process	Removes remaining impurities after alkali treatment and reorients the internal molecular structure	(i) Improves whiteness index (ii) The crystallinity increases higher than the alkali treatment (iii) Thermal stability and degradation temperature of the bacterial cellulose also increased after the bleaching process (iv) Increases the scope for other applications
3.	Colouration	Bacterial cellulose can be dyed using in-situ or ex-situ method by most of the researchers. The passive method of dyeing also showed a better fastness The addition of dye molecules in the bacterial cellulose reduces its intermolecular bonds	(i) Reduction on strength-related properties noted (ii) No much difference noted in the stiffness and moulding ability (iii) Comparable colour fastness with commercial material (iv) The in-situ method offers better colouration (colour strength) of bacterial cellulose with lesser process sequence

3.9 Drying Characteristics of Bacterial Cellulose

While discussing the practical application from common apparel or textile to technical applications like wound dressing, the material property at the long storage is important. For that, the material should be on a stable state in the atmospheric condition. The properties of just fermented or wet bacterial cellulose are well explained and analysed in recent years. However, in application point, drying of bacterial cellulose is identified as one of its important disadvantages as of now. The well-claimed property of the bacterial cellulose mat is its unique three-dimensional interconnected structures with micropore in it. The stability of the structure and application ability rises due to its randomly oriented fibre arrangement in its network of structure. This arrangement along with nano-sized fibre diameter is responsible for its exceptional properties in the wet state. However, when the material is dried, it loses its flexibility, the 3D nanofibrillar arrangement collapses and the pores in its structure reduce significantly. This change in structure highly affects its strength, water management and permeability-related properties in the application viewpoint [33].

Due to this structural change, the bacterial cellulose material's dehydration, reswelling of water, air permeability and water absorption reduce drastically. This reduces the interest of the material in different fields like fashion, biomedical, etc. Further, it is also reported that the type of drying method also has a significant influence on the morphological and structural aspects of the bacterial cellulose. Various drying methods adapted by different researchers are illustrated in Fig. 3.21.

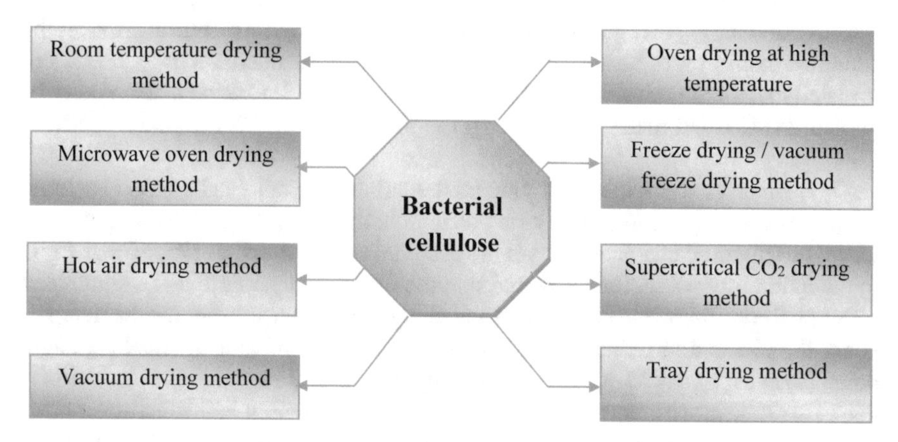

Fig. 3.21 Bacterial cellulose drying methods used by different researchers [Authors own illustration]

3.9.1 Oven-Drying Method

The oven-drying method is generally used to analyse the moisture content of a material. It is a thermo-gravimetric method of drying samples at a constant temperature for a defined time frame. The analysis is performed on the sample based on its weight loss upon drying by estimating the difference between the original and dried weights of the sample [71]. This method is generally used in bacterial cellulose drying process due to its availability and ease of use. Florentina Sederaviciute et al. developed cellulose using kombucha-based *G. bacter xylinus* in the tea medium and analysed the effect of drying on the mechanical properties of the dried cellulose film. The cultured cellulose film was neutralised with distilled water and squeezed with tissue paper. The bacterial cellulose is laid flat in the oven for different temperatures, namely 25, 50 and 75 °C. The results of the study revealed that the moisture loss in the wet bacterial cellulose was higher and quicker for the time in the case of higher temperature drying than lower temperature. The moisture loss was noted 91.8% for 50 °C and 88.3% for 75 °C.

Similarly, the celluloses dried at a lower temperature are softer than the dried one at a higher temperature. Concerning the tensile properties, the dried cellulose material showed a better tensile strength than the wet bacterial cellulose. Among the different temperatures analysed, higher breaking strength of 27.91 Mpa is noted for 25 °C dried bacterial cellulose. The lowest tensile strength value is 12.8 Mpa for 75 °C dried material. The strength value of 25 and 50 °C dried material did not vary much. Hence, the researcher reported the cellulosic materials are temperature sensitive and drying at higher temperature affects its internal structure and makes the cellulose matrix more brittle. The bursting strength test (biaxial tensile test) result revealed that maximum strength of 57.2 Mpa noted for 75 °C dried material and for 50 and 25 °C dried material the strength is 52.8 Mpa and 37.1 Mpa, respectively [72].

Domskiene et al. used oven-drying method to dry the kombucha-based bacterial cellulose and analysed its application potential in the apparel sector. The study evaluated the structural characteristics of the bacterial cellulose dried at 25, 50 and 75 °C in the laboratory oven. They reported that the drying temperature had a significant effect on the stiffness, tensile strength and elongation. Even at lower (25 °C) temperature, the elongation reduced 18.8% and significant increment in tensile strength noted (27.9 Mpa). At 75 °C, lowest breaking and elongation value is noted. This is due to the moisture evaporation and hydrogen bond formation in the internal molecular structure. For textile and apparel end use, the dimensional stability is one of the important parameters. Hence, they analysed and reported that the temperature did not have any influence on the bacterial cellulose dimensional change. But a maximum weight loss of 91.8% and thickness loss of 88.6% at 50 °C drying were noted followed by 25 °C. A diameter loss of 0.04% is noted in the case of 25 °C dried sample. Hence, the researcher concluded that the lower temperature drying (25 °C) preserves the porous nature and mechanical properties than higher temperature [73].

3.9.2 Freeze-Drying Method

Freeze drying is a process of dehydration of any material at a lower temperature. In this process, by using ice sublimation, the samples are used to freeze at a lower temperature to dehydrate its moisture content. This method is preferred over the conventional heat drying and evaporation method due to the high quality of the resultant sample. The main advantage of the process is the freeze-drying process mainly retains the shape of the sample over other methods. This is mainly due to the lower temperature processing [74]. Illa et al. produced bacterial cellulose using different bacterial strains (*Komagataeibacter hansenii* and from grape juice) and analysed its mechanical properties after drying with two different methods. The study used oven-drying and freeze-drying method after cultivation and purification process. The findings revealed that drying of bacterial cellulose greatly affects the crystallinity, fibre diameter, pore size, surface area, Young's modulus, tensile strength and strain. In comparison, the oven-dried samples possessed smaller crystal size but higher crystallinity than the freeze-dried sample (for grape-based bacteria 7.4% and *K. hansenii* 8.8%). When compared to the oven-dried sample, freeze-dried samples are noted less opaque and of higher thickness. This might be attributed to the heat evaporation of the oven-drying method, which forces the fibres to lose its moisture and makes it to move closer. This increases the orientation in the amorphous region in the oven-dried sample and increases crystalline nature [75]. In the case of freeze-dried sample, lower surface tension leads to higher thickness. In general, the freeze-dried samples are more porous and higher stretchable and can be used for moisture-related applications. Figure 3.22 represents the wet, oven-dried and freeze-dried bacterial cellulose samples.

This was further confirmed by the fibre diameter and pore size analysis. The average diameter of the oven-dried bacterial cellulose fibrils is 38.6 and 38.8 nm, respectively, for grape-based and *K. hansenii*. But in the case of the freeze-dried sample, the diameter of the fibre is noted 58.7 nm and 57.2 nm, respectively, in the same order. Similarly, an average pore size in the oven-dried sample is noted less than 100 nm but, in the freeze-dried sample, it is in the range of 50–350 nm.

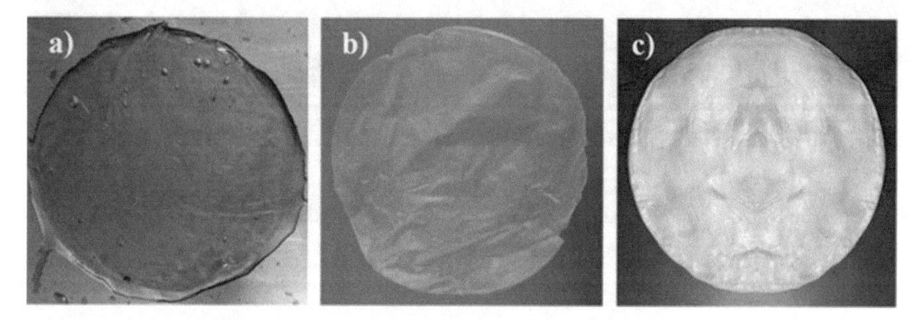

Fig. 3.22 Bacterial cellulose **a** before drying, **b** oven-dried and **c** freeze-dried [75] [Reprinted with Permission]

In common for both the species tested, the oven-dried sample showed a 65–70% reduction in pores. In the oven-dried method, the heating process rearranges the amorphous region and increases the hydrogen bonds in the section due to the moisture evaporation. This, in turn, collapses the structure, forms a denser arrangement of fibrils and reduces the pore size. Reduced fibre size and higher crystalline nature also contribute to the same. In the freeze-dried method due to the lower compressive tension on sublimation drying, there is a lesser amount of hydrogen bonding noted. This creates more of open structure, and lower surface tension forms a thicker but at the same time lower crystalline with a higher structurally stable sample. In the case of physical and mechanical test, compared to the freeze-dried sample oven-dried sample showed a superior strength, Young's modulus and lower strain due to its higher strength, closer and crystalline structure. The formation of hydrogen bonding is one of the most important reasons for all the differences noted with both the methods [75]. The scanning electron microscope picture of the oven-dried and freeze-dried samples is provided in Fig. 3.23.

Fig. 3.23 SEM micrographs of bacterial cellulose structure of **a** grape-based cellulose oven-dried, **b** freeze-dried, **c** *K. hansenii*-based oven-dried and **d** freeze-dried cellulose [75]. [Reprinted with Permission]

3.9.3 Tray Drying Method

The tray drying method is a conventional drying process that uses stacked trays to place the material inside an insulated drying chamber. The main difference in tray drying over oven drying is the heat in tray dryer is circulated using air inside the chamber [76]. NorhayatiPa'e et al. reported the drying nature of bacterial cellulose developed using *A. xylinum* in a coconut water medium (nata de coco). They compared the oven (at 80 °C), tray (at 80 °C) and freeze-drying (−50 °C) method on the swelling, mechanical and chemical properties of the developed cellulose. The findings revealed that the tray-dried samples shrunk more than the oven-dried and freeze-dried due to the better dehydration of moving air. On surface analysis, it is noted that the tray-dried samples are more compact and no pores noted than the other two methods. The oven-dried samples are rougher and stiffer with few pores, and the freeze-dried samples are with higher porosity. All the samples after treatment are evaluated for their crystalline nature and found very fewer differences among them. The crystallinity of the oven-dried, tray-dried and freeze-dried samples is found to be 87.80%, 87.01% and 88.90%, respectively.

In relation to crystalline behaviour, the author analysed the swelling behaviour of the dried cellulose. The swelling process represents the ability of the water to penetrate the structure of the cellulose after drying. The results represented that out of all the samples freeze-dried sample showed a higher swelling of 490% than the other structure. The swelling of the oven-dried sample (210%) and tray-dried samples (294%) is noted significantly lower due to its structural difference. Due to the heat application, the three-dimensional structure collapses and absorbs less water in both heat-dried methods. In specific, due to the complete collapse of structure and absence of pores the swelling is noted very less in the oven drying method. The second higher swelling % is noted with the tray-dried method as the structure was with few pores. In contrast, higher pores and thicker structure might be a reason for higher swelling of the freeze-dried sample [77]. Concerning the mechanical properties, the researcher reported a contradictive result to previous work [75] that the freeze-dried samples showed a higher strength followed by tray-dried and oven-dried [77].

In another study, the freeze-dried bacterial cellulose samples are compared with the oven-dried samples for its structural properties. In appearance, the oven-dried samples are transparent than freeze-dried and later had a higher thickness (627 ± 84 μm) than the oven-dried (12 ± 2 μm). The morphology analysis revealed that freeze-dried samples are more open and porous than the oven-dried samples, in which the structure was collapsed. With respect to thermal degradation, the type of drying techniques does not have any influence on it. In the chemical structure, general crystallinity does not alter with drying type. However, the researcher noted that there is a slight change in the proportions of cellulose I and II content while analysing crystallinity using peak height differences. Similarly, the FTIR analysis reported a formation of new crystal structure in the oven-dried sample than the freeze-dried sample. The amorphous region, 800–900 cm^{-1}, reported significant changes and represented a broadening of the peak as a representative of the more amorphous region in the freeze-dried sample. In contrast, the oven-dried sample showed an increase in the crystalline peak at 2900 cm^{-1}. This confirmed that the freeze-dried sample showed a higher cellulose Iα content than the oven-dried. This confirms the effect of the drying process on the internal structure of bacterial cellulose [78].

3.9.4 Room Temperature Drying Method

Room temperature drying is one of the simplest and cost-effective methods of drying bacterial cellulose. In this method, bacterial cellulose is placed in the open air at room temperature to lose its moisture till it attains a constant weight. The moisture loss is evaluated by calculating the difference between the initial and final or dried mass of the bacterial cellulose. In room temperature drying, the structural modification in terms of porosity and density is usually expected due to the compressive force created by the atmospheric pressure. In research work, the effect of room temperature was compared with freeze-dried sample for its characterisation. The crystallinity analysis of the research revealed that the room temperature and freeze-dried sample showed a similar crystallinity percentage (71 and 76%). Out of both the methods used, room temperature drying had the least film thickness and higher fibre density. This is due to the heat application in room temperature than the freeze-dried method, where in freeze drying, the temperature is brought down to −80 °C [79].

Concerning the porosity, there is no difference noted with the freeze drying and room temperature drying. Both the methods offered a porosity of 94% and 95%, respectively. In water absorption capacity, the researchers reported a major influence of drying technique and found that water absorption capacity of the room temperature-dried cellulose is significantly higher than the freeze-drying method. Upon consecutive repetitive experiment, the results found the same. This was mainly due to the bundling of fibrils inside the microstructure during the drying process. Nearly 85% of fibres are bundled together after both the drying processes [10]. Other researchers developed *A. xylinum*-based cellulose using various media, namely HS, HS with mannitol, high fructose corn syrup, high fructose corn syrup mannitol, molasses and molasses mannitol. The resultant cellulose is dried at room temperature, and the results are compared with freeze-dried samples. Out of the tests performed, two findings are noted as important. First, concerning the breaking strength, irrespective of the media type used, the freeze-dried sample had the highest strength (in N) value than the air-dried samples. Secondly, the breaking strength values of the bacterial cellulose samples produced from mannitol medium were always higher irrespective of the drying method used. In addition to that, the air-dried samples showed a higher breaking elongation and abrasion resistance value than the freeze-dried sample. In the mannitol medium, the effect is still noted higher [80].

3.9.5 Supercritical CO_2 Drying (SCD)

Supercritical CO_2 drying is a process of removing moisture content from a product in a controlled manner. The process is also referred to as critical point drying [81]. The process is very similar to the air-drying process but instead of air, in this process, carbon dioxide is used at 74 bar pressure and temperature above 31 °C. At this stage, the carbon dioxide gas can easily penetrate the samples like gas with a density of the liquid. This is called as the supercritical state of CO_2. This process is very faster than any other drying method, and the used CO_2 can be recollected and used again; hence, this process is known for its lesser energy consumption, and efficient and sustainable nature [82]. This method is also used by some researchers to analyse the drying characteristics of bacterial cellulose, and they compared it with the freeze-drying and room temperature drying method. The research found that the supercritical temperature drying had a significant impact on the bacterial strain tested. The researcher evaluated two different strains, namely *Gluconacetobacter xylinum* (GX) and *Gluconacetobacter europaeus* (GE). The crystallinity analysis results reported a higher percentage for SCD method than the other two methods for both strains selected. As the samples are obtained from the same source, the increase in crystallinity might be attributed to the SCD method used.

Further, the thickness of the SCD samples is noted higher than the room temperature and freeze-dried samples. This might be attributed to the elimination of solvent meniscus at supercritical solvent evacuation. In the subsequent analysis, the water absorption capacity of the SCD samples is very high up to 110 times of its weight whereas it is 20 times for room temperature-dried and freeze-dried sample. To support this finding, the scanning electron micrograph image revealed that the fibres bundled together in all the methods used. However, in SCD sample (38%) less amount of fibres only bundled when compared to the room temperature-dried and freeze-dried sample (85%). And hence, more porous structure is noted in the case of SCD sample than other two methods. This bundling of fibres also altered the surface roughness in the other two methods and hence there may be a difference in water absorption capacity of the room temperature-dried and freeze-dried samples than SCD method [79]. The researcher also analysed other structural and mechanical properties, and then compared among the three methods as listed in Table 3.4.

3.9.6 Hot Air-Drying Method

It is another common method used in the drying of vegetables and fruits. In this process, heat is applied by the hot air on the product. Thus, it evaporates the moisture by the convection process of the hot air. The major disadvantage of the process is it requires comparatively a longer time to dry the product. The direct heat application sometimes degrades the quality of the product dried under this method [83]. *G. xylinum*-based bacterial cellulose material was analysed by hot air-drying method

Table 3.4 Effect of different room temperature, freeze and supercritical drying methods on properties of bacterial cellulose [79] [Reprinted with Permission]

Drying method	Room temperature drying		Freeze drying		Supercritical drying	
Strain	GE	GX	GE	GX	GE	GX
Structural properties						
Film thickness (μm)	24 ± 2	37 ± 2	34 ± 2	40 ± 4	58 ± 3	67 ± 2
Density (g/mL)	0.08 ± 0.01	0.59 ± 0.03	0.06 ± 0.01	0.32 ± 0.02	0.05 ± 0.01	0.16 ± 0.01
Total porosity (%)	94 ± 2	60 ± 2	95 ± 2	79 ± 2	96 ± 2	89 ± 2
Fibre diameter (nm)	17 ± 2	19 ± 3	20 ± 4	19 ± 3	16 ± 2	20 ± 4
Water absorption capacity (WAC)						
1st cycle	39.5	37.3	15.8	15.8	109.4	66.6
After 3 cycles	15.3	14.2	92.3	50.2	198	19.4
Decrease (%)	61	62	4	68	82	71
Mechanical properties						
Penetration depth at 0.4 mN load (μm)	3.9 ± 0.4	1.1 ± 0.1	4.1 ± 0.4	1.4 ± 0.3	3.2 ± 0.5	4.3 ± 1.5
Young modulus ('E') (MPa)	198 ± 46	659 ± 85	204 ± 46	601 ± 155	238 ± 25	198 ± 34
Hardness ('H') (MPa)	34 ± 22	39 ± 9	20 ± 8	26 ± 12	20 ± 5	19 ± 21
EP (%)	59 ± 7	39 ± 3	53 ± 4	33 ± 3	52 ± 5	48 ± 13

by Chuanjie Zhang et al. The researcher analysed the effect of hot air-drying, freeze-drying method on the chemical and morphological properties of the developed bacterial cellulose. They reported that the hot air method significantly reduced the thickness of the material. In the morphological analysis, they concluded that the hot-dried samples are very compact and tight without pores, similar to oven-dried one. The freeze-dried material is highly layered and more porous than the hot air-dried. Compared to oven drying, the air-drying method is more spontaneous and so it loses its moisture very quickly due to the flow of air. In the case of the crystallinity, due to higher dehydration and creation of compact structure, the hot air-dried samples showed a higher crystallinity of 73.9% than the freeze-dried one of 58.2%. The

FTIR analysis of the sample confirmed the loose structure of the freeze-dried sample over hot air-dried one, by showing a deviation in –OH stretching of molecules at 3345 cm^{-1}. This represents the weaker hydrogen bonds in the freeze-dried sample than the hot air-dried samples of bacterial cellulose. In the mechanical properties, hot-dried bacterial cellulose showed a higher breaking strength (28.5 Mpa) than the freeze-dried sample (12.4 Mpa). Young's modulus of the hot air-dried samples is far high (13.02 Gpa) than the freeze-dried one (1.5 Gpa). These findings suggested that when the mechanical properties are considered the use of hot air drying is better than a freeze-dried sample [84].

3.9.7 Vacuum Drying Method

In vacuum drying method, temperature-sensitive products are dried at a lower temperature. In this process, the sample is kept in an enclosed container and the air pressure is reduced by an external vacuum pump. The method generally dries the product at 25 °C, three times faster than hot air drying at 30 °C and humidity of 50%. This process reduces the vapour pressure of the water and boils it. The decrease in boiling point increases the evaporation of moisture or drying rate of the sample [85]. In a comparative study, the researcher evaluated the effect of a vacuum drying method with oven-dried and freeze-drying methods. They reported a significant change with the structure of the dried bacterial cellulose. In the morphological analysis, the vacuum-dried sample showed a medium entanglement of fibres with lots of cavities than the oven-dried. Though oven-dried and vacuum-dried samples showed a compact structure than the freeze-dried, the surface roughness and cavities are noted higher in the case of vacuum-assisted drying [84]. There was no much difference noted in the IR spectra of the oven-dried sample and vacuum-dried sample but much variations in the case of the freeze-dried sample as discussed in the earlier section. However, the crystallinity percentage of 63.8% noted for vacuum-dried method is lower than the oven-dried. The oven-dried samples are more compact and provided a higher breaking strength than the vacuum-dried sample (22.6 Mpa). The freeze-dried sample possessed a lower strength than the vacuum-dried sample. Young's modulus value of the vacuum-dried sample also took a place in between oven-dried and freeze-dried samples [84].

3.9.8 Microwave Oven Method

For the drying of bacterial cellulose, few researchers also used a microwave oven, which we commonly use for cooking food. It is one of the common methods used in the food industry to dry vegetables and fruits. It uses high-frequency electromagnetic waves with a frequency range of 300 MHz to 300 GHz. It is also considered to be a moderate to low energy consuming method of drying material, and this volumetric method reduces the processing time. In this process, the ability of the material to absorb the waves modifies it as heat. When the microwave penetrates into the material, the formation of heat energy due to the conversion evaporates the moisture content of the material [86]. The researchers reported that the water molecules present in the cellulose show higher polar nature and so it aids more transformation of heat from microwave energy than the convection heating process [87]. The harvested, purified bacterial cellulose is dried with microwave oven and compared with the hot air oven-dried bacterial cellulose of the same quality. The microwave drying made the bacterial cellulose surface very smooth. The main reason is the proper distribution of the heat throughout the structure by water molecule vibration. Over hot air-drying method, microwave drying targeted the sample in a bulk over the surface heating and in less time. The hot air-drying method created wrinkles on the surface of bacterial cellulose due to the faster water evaporation.

The drying time of the sample is one of the major advantages of the microwave method, as it took 3–5 min to dry the bacterial cellulose, and in the case of hot air oven drying it is 3–5 h. Further in chemical analysis, the researcher reported more similarities between the drying methods. However, in both samples compared to wet sample more amount of hydrogen bond between the cellulose chains is represented by 3342 cm^{-1} peak. XRD analysis also produced similar spectra, representing similar structural changes in both methods. The crystallinity of the air dry oven method is 88.5% and microwave method is 84%. The difference might be longer drying time of the hot air oven method than the microwave drying. This time might have allowed the interfibril molecules to move together and form more hydrogen bond than the microwave drying. Concerning the swelling properties, the microwave-dried samples showed higher swelling percentage (264%) than hot air oven-dried (242%). This is because the microwave-dried samples had lower crystallinity and less number of hydrogen bonding; this indicates the amorphous region of the sample. The tensile analysis showed a higher strength of 474.34 Mpa than the hot air-dried one (448.86 Mpa). However, the microwave-dried showed a lower elongation than hot air oven-dried. This might be attributed to the surface wrinkles of the hot air-dried samples over microwave-dried one [87]. Table 3.5 compares the main pros and cons of the various methods discussed in this section as consolidated.

3.10 Functionalisation with Plasticisers

Functionalisation of textile material with different finishing and coating substances is more common for their application requirement. Similar attempts were also made on bacterial cellulose non-woven mat to enhance its application potential. The functionalisation process was performed in two cases, either to include some of the functional characteristics like antimicrobial, etc., or to enhance some common properties like stretchability, hydrophobicity, etc. This process can be performed in two major methods: (i) inclusion of water-soluble plasticisers or other chemicals in the bacterial cellulose production medium itself during the synthesis, known as in situ,

Table 3.5 Pros and cons of various drying methods used in the bacterial cellulose drying process

S. no.	Drying methods	Advantages	Disadvantages
1.	Oven drying	• Created higher crystallinity with lower crystal size • Lower fibre diameter • Higher Young's modulus • Higher tensile strength • Higher surface tension • More opaque structure	• Lower pore size, hence permeability properties will be affected including moisture handling ability • Lower structural integrity. Hence, shrinkage cannot be controlled • Hard, rough and stiff structure makes it difficult for textile and apparel applications
2.	Freeze drying	• More open structure and higher pore volume • Due to sublimation dehydration, structural integrity not affected • No shrinkage in the size and shape • Higher thickness and less transparent appearance • Smooth surface	• Reduced mechanical properties than oven-dried material • Lower surface tension
3.	Tray drying	• Higher crystallinity due to higher weight loss and more compact structure • Very less water swelling nature	• Higher weight loss and structural collapse than oven drying and freeze drying • No pores in the structure after drying
4.	Room temperature drying	• Higher material thickness • Higher crystalline nature than freeze • Higher water absorption capacity • Higher tensile strain and abrasion resistance percentage	• Lower crystalline than the higher temperature drying techniques • Longer drying time • Dimensional shrinkage
5.	Supercritical CO_2 drying	• Higher crystallinity than room temperature drying and freeze drying • Higher water absorbing capacity • Highly porous structure than other drying methods • Higher Young's modulus	• Higher thickness of the material • The lower hardness value of the material • Lower film density than other material

(continued)

Table 3.5 (continued)

S. no.	Drying methods	Advantages	Disadvantages
6.	Hot air drying	• Similar to tray and oven drying creates higher crystalline nature • Higher Young's modulus and breaking strength • Compact and bundled fibril arrangement • Lower thickness than freeze-dried samples	• Less porous in nature • Lower tensile strain percentage • Poor water absorbency or moisture-related properties
7.	Vacuum freeze drying	• More compact structure than the freeze-dried sample • Higher tensile strength and Young's modulus than the freeze-dried sample	• More surface imperfections noted and so higher roughness • Lower crystallinity percentage than the oven-dried sample • Lower mechanical properties than oven-dried sample
8.	Microwave drying method	• Shorter drying time • Better heat distribution leads to the smooth surface structure • No changes in general chemical structure while comparing with hot air oven drying • Higher tensile strength value over the oven-dried sample • Higher swelling percentage	• Lower crystallinity percentage than the oven-dried sample • Lower elongation at break percentage

and (ii) treatment of the chemicals or plasticisers on the bacterial cellulose after the synthesis through suitable chemical reaction or entrapment of the component in the structure called as ex situ.

The presence of a large number of hydrogen groups in the cellulose matrix encourages the chemical modifications. Concerning bacterial cellulose, more of the plasticisation is experimented than the functionalisation. Generally, drying bacterial cellulose makes it stiffer, brittle and lowers the breaking elongation percentage with different drying methods. However, for textile and fashion application the material still needs to be flexible. To enhance the flexibility and other functional properties, several researchers functionalised the bacterial cellulose mat with different plasticisers and softeners. This section consolidated the various attempts made in the literature to functionalise the dried bacterial cellulose mat.

3.10.1 Use of Plasticisers (Ex-Situ Applications)

The special feature of the bacterial cellulose is the unique three-dimensional interconnected porous structure of the developed matrix. The nano-sized fibrils are also another important factor compared to the plant cellulose that offers a great amount of surface area and that provides its unusual properties. Though the bacterial cellulose has numerous potential properties in the wet state, they are not stable over different temperatures. Hence, the drying and stabilisation of the structure are important. The different modes of drying evaporate the internal moisture content rapidly, lead to shrinkage and loss of flexibility, and create stiffness and other permeability-related issues as discussed in the previous section. Irrespective of the method used, drying generally increases the crystallinity of the bacterial cellulose and reduces the flexibility. However, it alters the mechanical and other properties related to the chemical structure. The most commonly adopted method to keep unaltered 3D structure is the freeze-drying method. But, the addition of plasticiser will decrease the intermolecular forces between the structures and increase the structural mobility of the polymer chain. This facilitates the bacterial cellulose to be elastic and less brittle by reducing the glass transition temperature of the material [88]. The major concern about the ex-situ modification is the material or chemical that is used for the modification or functionalisation of the bacterial cellulose should be of nano-sized materials. That is the only possibility to distribute the chemical into the bulk structure of the cellulose matrix.

In a study, researchers used *G. bacter xylinus*-based bacterial cellulose for the plasticising process. After fermentation and purification with NaOH, the cellulose mat is treated with glycerol and polyethylene glycol. The formation of additional hydrogen bond on the drying activity is the main reason for the restricted movement of the internal molecules. Hence, when a plasticiser component is added in the bacterial cellulose molecule, it interacts with adjacent fibres and forms hydrogen bonds with each other. This process increases the interspace between the fibres and reduces the fibril-to-fibril hydrogen bonding to a greater extent. This process ultimately increases the interspace and so the fibre movement. This is the fundamental mechanism behind the use of any plasticiser in bacterial cellulose. The researcher used glycerol and polyethylene glycol as a plasticiser, and the mechanism is illustrated in Fig. 3.24 [89]. The developed bacterial cellulose is placed in the plasticiser solution with different concentrations for 24 h and naturally air-dried. The morphological analysis reported a very less stacking of nanofibrils in the scanning micrograph. The addition of plasticiser increased the pore and cracks of the structure. This increment further increased the surface roughness of the plasticiser-treated bacterial cellulose. Thus, the surface roughness increases with the plasticiser concentration.

In the case of contact angle analysis, the addition of glycerol increased the hydrophobic nature of the cellulose; however, the contact angle of the 2% glycerol-treated sample is less compared to the 1% glycerol-treated sample. In the case of polyethylene glycol, the same trend is noted. Though the plasticisers have lower

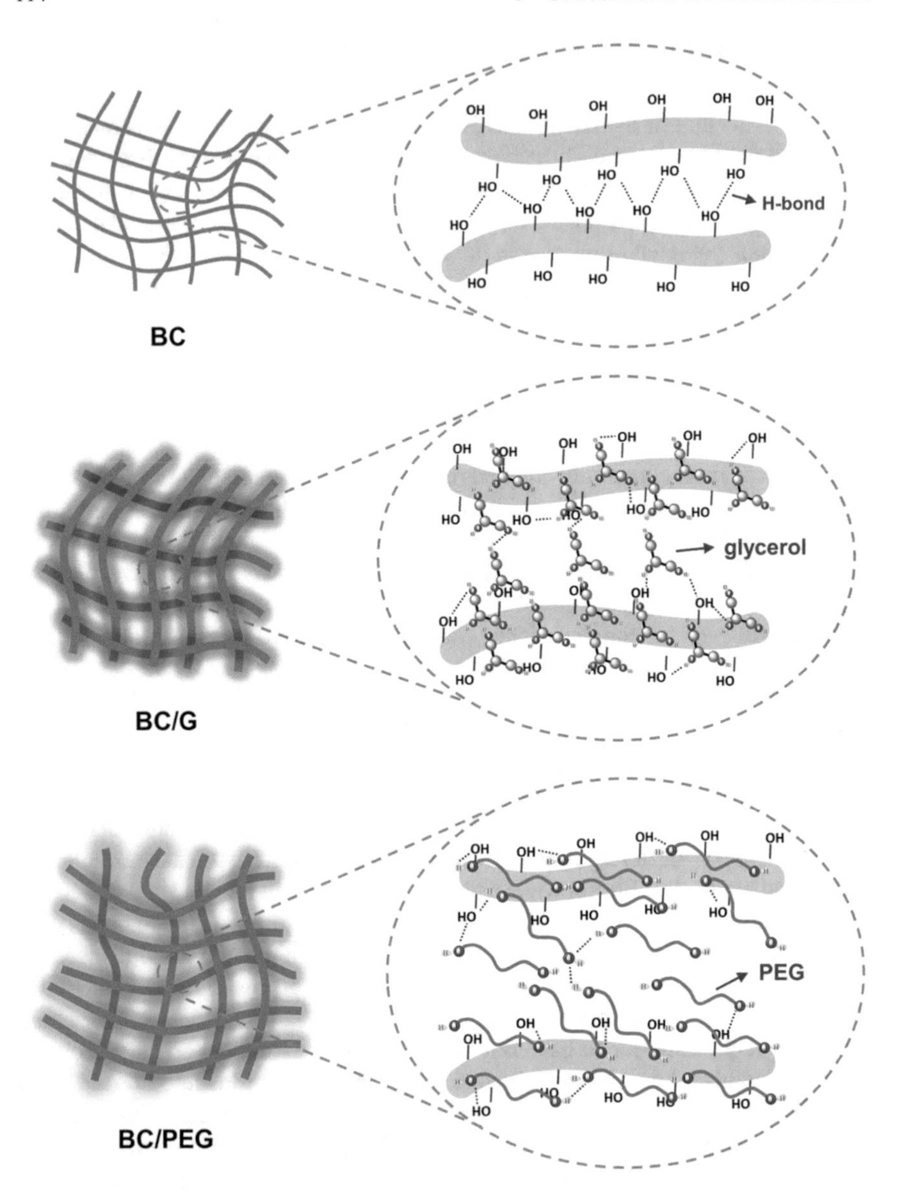

Fig. 3.24 Plasticiser interaction mechanism with bacterial cellulose between glycerol and polyethylene glycol [89] [Reprinted with Permission]

hydroxyl groups, an increase in the concentration might have improved the reduction in contact angle. Similarly, the addition of plasticiser reduced the tensile strength of the material to a greater extent and increased the breaking elongation for both the plasticisers used. The water absorption capacity, water retention capacity and water vapour transfer rate of the samples were also increased with the concentration of the plasticiser. These changes might be attributed to the increased interfibre space, and less amount of hydrogen bonding facilitated the changes in the property [89].

Komagataei bacterxylinus-based bacterial cellulose composite material was developed using carboxymethyl cellulose and hydroxyethyl cellulose in it. The resultant products were treated with glycerol for plasticising effect and analysed for its essential properties. Initially, by focusing on the wound exudate absorption capacity both the composites are treated with glycerol concentration ranging from 0.1 to 2.5%. The results indicated that the concentration of glycerol had a considerable impact on the property. A maximum of 19 g exudates is absorbed by the composite of 1 g dry weight in the 0.5–2.5% of glycerol concentration. The researcher noted that the maximum absorption capacity is noted in the glycerol concentration range of 0.5–1% (approximately 16–17 g per 1 g dry weight).

A maximum of 54% exudate absorption noted at higher percentage (2.5%) of the glycerol-treated sample but in the case of lower concentration less than 1%, a maximum of 60–85% exudates absorbed in 24 h immersion. The researcher also proposed that the addition of glycerol modifies the rehydration properties of bacterial cellulose and composites. The rehydration capacity of the normal bacterial cellulose is noted as 2.4% and noted as 4.6 and 9.1% for carboxymethyl cellulose and or hydroxyethyl cellulose composite in 24 h. The addition of glycerol has a higher impact on the rehydration capacity of the bacterial cellulose and its composites. The researcher analysed the different concentrations ranging from 0 to 10% of glycerol on native bacterial cellulose and developed composites. They reported that irrespective of glycerol concentration and treatment time the bacterial cellulose and carboxymethyl cellulose composite had higher rehydration capacity than the other samples. The rehydration capacity increases significantly after 2.5% and reaches a maximum at 10% glycerol concentration. Interestingly, on mechanical property analysis, the addition of glycerol showed an increase in breaking stress and Young's modulus value. However, the increment is very higher in percentage in the case of 2.5% glycerol concentration than the 10% glycerol. Based on these analyses, the findings suggested the use of glycerol-treated bacterial cellulose composite as a potential material for wound dressing application [90].

A different group of researchers used glycerol as a plasticiser for bacterial cellulose and developed polyvinyl alcohol composite film from *Gluconacetobacter xylinus*. The researchers analysed the toughness, puncture, swelling, solubility and optical properties by considering the food packaging application. The developed cellulose is purified and treated with glycerol for 2 h and dried at 24 °C for 48 h. The morphological structure showed that the addition of polyvinyl alcohol increased the density of the cellulose and smoothened its surface with higher concentration. Addition of glycerol did not have any significant influence on the film thickness as polyvinyl alcohol. In the case of mechanical properties, the addition of glycerol in the composite reduced the toughness of the film significantly. The addition of glycerol did not have any negative effect on the bursting strength, but it increased the elasticity of the film due to the increased lubrication among the molecular structure. Similarly, the swelling and soluble capacity of the bacterial cellulose composite with glycerol reduced significantly than the untreated composite and native cellulose [91]. Lucia Indrarti et al. developed a bacterial cellulose composite film using glycerol, sorbitol and carboxymethyl cellulose. The purchased cellulose pellicle is blended with glycerol, sorbitol and carboxymethyl cellulose to develope individual composites. The solubility analysis revealed that the addition of glycerol and sorbitol increased the solubility of the composite film, and in the case of carboxymethyl cellulose composite, the solubility is reduced to a great extent. In the case of tensile strength, the addition of glycerol and sorbitol significantly reduced the strength from 84.1 Mpa to 66.7 Mpa and 79.4 Mpa, respectively. But for the carboxymethyl cellulose composite, the tensile strength is noted as 180 Mpa that is very higher than the control bacterial cellulose. The difference is mainly attributed to the plasticising effect of plasticisers used. When the plasticiser is used, it reduces the intermolecular bonds of the cellulose and increases the bonds between cellulose and plasticiser. Thus, it increases the flexibility of the film by increasing the mobility of intermolecular chain. Hence, the tensile strength reduces and so the extension to break increased for the plasticised composites. But at the same time, carboxymethyl in cellulose acts as a cross-linking agent and creates more hydrogen bonds between the cellulose. Thus, it increases the tensile strength of the bacterial cellulose composite. Based on the elastic modulus and extension to break percentage, the authors also suggested that the carboxymethyl cellulose also acted as a plasticiser in the developed composite [92].

Rohaeti et al. developed bacterial cellulose from rice waste water and developed composite with glycerol and chitosan. The researchers evaluated the mechanical and chemical properties of developed bacterial cellulose composite. On characterisation, the researcher reported that the water absorption capacity of the native bacterial cellulose and glycerol-treated composite was higher than the chitosan-added composite. This might be due to the higher space in the internal molecular structure in the previous cases over chitosan composite. In the morphological analysis, the glycerol treatment has covered the fibrils completely, penetrated into the fibre and developed a bulk structure than the native and chitosan-based bacterial cellulose. In mechanical property analysis, the tensile strength results showed a reduction with glycerol-treated cellulose than native and chitosan composite. However, a higher elastic nature is noted with glycerol-treated bacterial cellulose. This is mainly due to the reduction in the intermolecular forces by the addition of glycerol in the cellulose matrix. To confirm this, a lower crystallinity percentage of 47% is noted for the glycerol-treated bacterial cellulose than others. To add functionality, both the glycerol and chitosan-added cellulose showed an antimicrobial activity of 11.6 and 11.8 mm [93]. Other researchers developed bacterial cellulose from coconut water using *A. xylinum.* The developed bacterial cellulose is treated with glycerol, and its characteristics were analysed [94].

Fernandes et al. used a fabric softener, polydimethylsiloxane as plasticiser and perfluorocarbon as a hydrophobiser in the bacterial cellulose matrix to increase its flexibility. In their research, morphological analysis results reported an increment in the thickness of the developed bacterial cellulose mat with the addition of softener. At higher concentration, the thickness of the cellulose matrix increased for both the

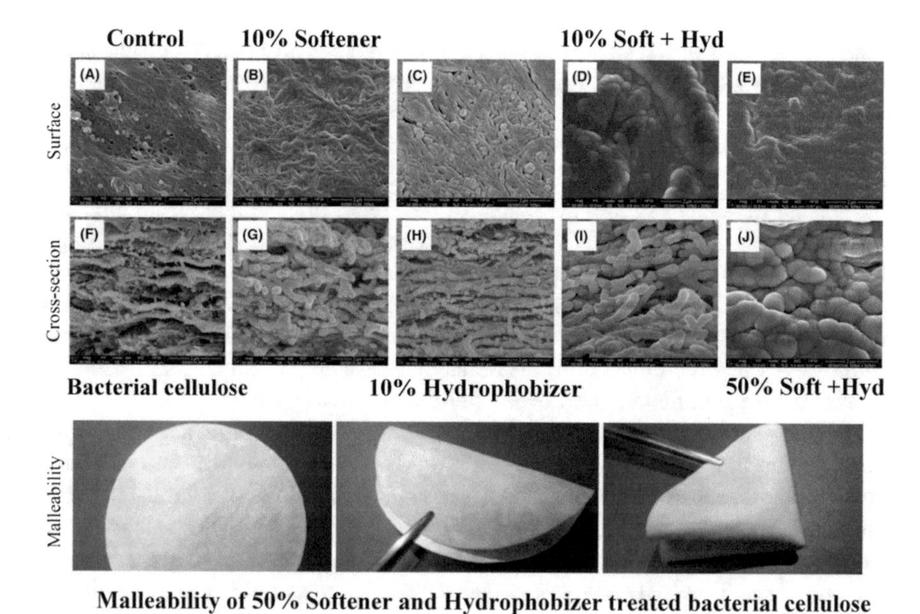

Fig. 3.25 SEM images of the softener and hydrophobiser-treated bacterial cellulose [33] [Reprinted with Permission]

treatments. The scanning micrograph images of the treated bacterial cellulose are provided in Fig. 3.25 for better understanding [33].

The atomic force microscope analysis revealed that the addition of softener significantly reduced the surface roughness than the native cellulose and hydrophobiser-treated cellulose. This was attributed to the filling effect of softener on the bacterial cellulose over the hydrophobiser. The changes in surface roughness and introduction of polymers into bacterial cellulose change the wettability and contact angle of the treated cellulose. The lowest contact angle is noted for the control sample, and other samples showed a higher contact angle. For water vapour permeability of the sample, the addition of the polymers in the cellulose matrix reduces the permeability due to the increment in the thickness and filling of polymer in the intermolecular structure. In mechanical property, the addition of softener in cellulose structure initially increased the tensile strength of the film at 1% concentration and then reduced at 10, 25 and 50%. Similarly, in the case of hydrophobiser an increment is noted up to 10% and in combined treatment also the reduction point is noted as 10%. Based on the mechanical and permeability properties, the researcher suggested the developed bacterial cellulose composite as an alternative for leathers in shoe application [33].

Jennifer analysed the effect of glycerol treatment on the bacterial cellulose developed from different carbon sources and dried using air-drying and freeze-drying method. Overall, all the mechanical properties of the developed bacterial cellulose increased with the glycerol-treated samples than the non-glycerol-treated. In particular, the bacterial cellulose produced in the presence of mannitol medium produced bacterial cellulose with higher tensile strength in both glycerol treated and untreated. Among the two, the treated samples showed a significant improvement in the tensile strength. Out of all the media, a maximum tensile strength is noted for mannitol-added HS medium over other media. In this medium, air-dried samples showed a higher tensile strength of 15.18 N and the freeze-dried sample had a value of 13.78 N. The second maximum strength is noted for the freeze-dried sample from molasses and mannitol medium (11.61 N). In case of breaking extension, a maximum elongation of 11.61 mm is noted for the molasses mannitol freeze-dried sample followed by 11.56 mm in freeze-dried of the mannitol-added HS medium. A similar extension of 11.06 mm is noted in the air-dried mannitol and molasses medium-based bacterial cellulose. As like the elongation, the abrasion resistance of the bacterial cellulose samples is also significantly improved after the glycerol treatment [80].

Other researchers tried to resolve this issue with the use of citric acid as a cross-linking agent in the bacterial cellulose matrix. Based on the use of citric acid as a cross-linking agent in the plant cellulose structures, the authors tried it for bacterial cellulose also to enhance the rehydration properties. The findings of the study confirmed the cross-linking of the cellulose fibrils with citric acid by using ATR-FTIR analysis. The XRD analysis revealed that the cross-linking process did not affect the crystallinity of the treated sample over the native cellulose. The SEM image analysis revealed that the surface and cross section of the purified cellulose are very similar to the citric acid cross-linked bacterial cellulose. The cross-linked material also possesses the same amount of surface pores as similar to the purified control and freeze-dried samples. The surface and cross sections of the cellulose structures are provided in Fig. 3.26 [95].

The cross-linking prevents the structural collapse of the cellulose and maintains the pores and structural integrity. However, the mechanical property analysis results showed a greater reduction in the tensile strength noted 4–5 times lesser, from 30 N (purified untreated) to 8 N for cross-linked bacterial cellulose. A similar effect is also noted in the elongation percentage from 11 to 2% after treatment. This is mainly due to the reduction in the mobility of the cellulose due to the cross-linking reaction of the citric acid. This was again confirmed by the increment in the thickness of the material (137 times) than the control. Concerning the water absorbency, the cross-linked sample absorbed 3 times higher water than the control sample. This effect can be expected due to the higher porous nature and thickness of the material. Similarly, a higher swelling rate is also noted for the cross-linking sample. The researcher proposed the cross-linking method is one of the best methods to increase the rehydration of dried bacterial cellulose. This will aid the performance in the applications like wound dressing and other applications which requires absorption. Figure 3.27 represents the purified untreated bacterial cellulose and cross-linked cellulose with citric acid [95].

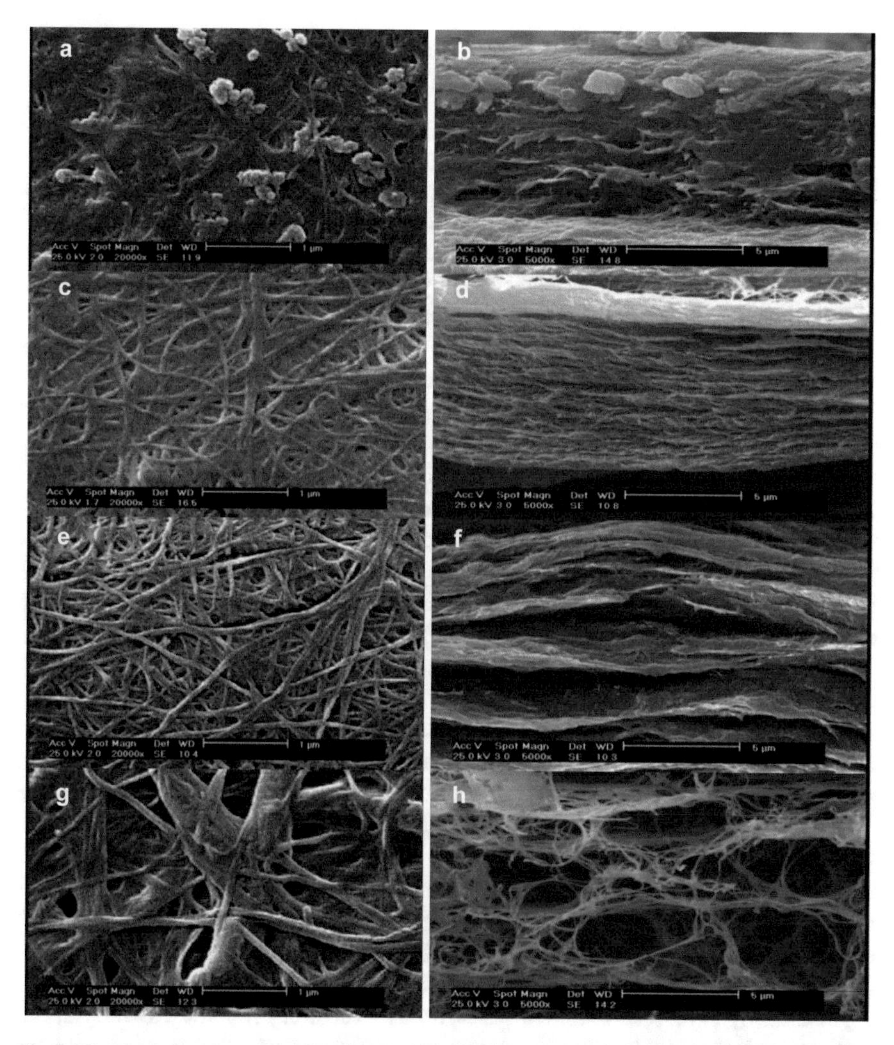

Fig. 3.26 Top and cross-sectional images of **a**, **b** unpurified, **c**, **d** purified, **e**, **f** citric acid cross-linked and **g**, **h** freeze-dried bacterial cellulose samples [95] [Reprinted with Permission]

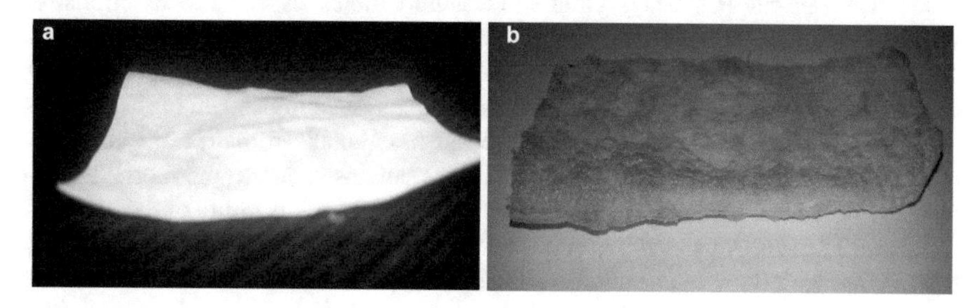

Fig. 3.27 **a** Purified bacterial cellulose (untreated) and **b** citric acid cross-linked bacterial cellulose matrix [95] [Reprinted with Permission]

3.10.2 Use of Functional Components (As Composites)

Similar to plasticisers, several other functional chemicals were also used to impart different functional properties and some essential properties to the bacterial cellulose. Lauryl gallate, the n-docosyl ester of gallic acid and a functional polymer are commonly used for cellulosic fibre modification. The treatment creates an enzymatic oligomerisation and improves the fibre properties significantly. In this aspect, a recent study analysed the use of the lauryl gallate oligomers in the stabilisation of bacterial cellulose and it improves the mechanical and chemical properties after drying. In this process, the developed bacterial cellulose is treated in the oligomer solution after the swelling process. The cellulose matrix impregnated in the oligomer for 12 h at 50° C. On the mechanical property analysis, the addition of lauryl gallate converted the cellulose non-woven as hydrophobic material. The addition of lauryl gallate reacts with free hydroxyl groups and reduces the moisture absorption properties. The water contact angle of the treated bacterial cellulose increased to 118° for highest concentration than the untreated cellulose. The physical entrapment of the lauryl gallate was confirmed by surface atomic composition with the increment in the N1s percentage of the treated bacterial cellulose over the C1s and O1s in the untreated one [19].

In addition to that, the crystallinity analysis showed an increase in the crystallinity from 73.8 to 79.3%, by representing the reason for the reduction in moisture uptake ability. This is attributed by the entrapment of the lauryl gallate. The researchers also reported that the treated non-woven fabrics possessed a higher tensile strength (88%) compared to the untreated one in both dry and wet states. The treatment reduced the dimensional shrinkage of the bacterial cellulose after wetting, and it is reduced from 40.5 to 19%. Even after a prolonged immersion in water, the treated fabric showed better dimensional stability. The author reported the lauryl gallate functionalisation can be effectively used to increase the mechanical properties of the bacterial cellulose non-woven which increases its application potential in many disciplines [19]. Bacterial cellulose has been reported as an ideal material for wound dressing application by several researchers based on its structural properties. However, the material itself does not have any antibacterial activity by itself. Hence, it is important to functionalise the material to meet the requirements. In this basis, few researchers used cerium nitrate ($Ce(NO_3)_3$) and silver nanoparticles (AgNPs) to functionalise bacterial cellulose by using oven-drying and autoclave method. The morphological analysis of $Ce(NO_3)_3$ and AgNPs-added bacterial cellulose showed a different colour than native cellulose. Along with that, native bacterial cellulose had highly porous and organised three-dimensional structure but functionalised matrix did not show any fibrils and the structure also appeared to be collapsed due to the thermal treatment. This, in turn, imparted a serious effect of the rehydration ability of the bacterial cellulose. The functionalised bacterial cellulose showed a poor rehydration capacity than the native cellulose. The addition of $Ce(NO_3)_3$ and AgNPs in the matrix reacted with most of the available hydroxyl groups in the cellulose. This might be the reason for the reduction in the rehydration capacity of the treated bacterial cellulose over the native one.

The SEM image also showed an increment in the fibre diameter upon the treatment of $Ce(NO_3)_3$ and AgNPs. This confirms the coating of the $Ce(NO_3)_3$ and AgNPs on the bacterial cellulose. The XRD analysis showed a 5.4% reduction in the crystallinity percentage for the $Ce(NO_3)_3$ and AgNPs-treated bacterial cellulose. The changes in the crystallinity of the functionalised bacterial cellulose are due to the addition of metal elements and also the reduction in hydrogen bonds in between the intermolecular structures of the bacterial cellulose. The functionalised cellulose matrixes showed a higher antimicrobial activity for the oven-dried sample than the autoclaved one. A maximum of 13 and 17 mm zone of inhibition is noted against the *Pseudomonas aeruginosa* and *Staphylococcus aureus* for the oven-dried sample. The open structure and higher impregnation of the chemical in the autoclaved sample are the reasons for its higher leachability over oven-dried sample [96]. Numerous attempts were made by several research workers to functionalise the bacterial cellulose and use it in the biomedical field. As the general interaction and mechanism remain similar (ex situ), the influence of the particular chemical on the mechanical and chemical properties of the cellulose is subjective. Table 3.6 consolidates the various recent research works on bacterial cellulose functionalisation for specific end use.

3.11 Summary

The main idea of the chapter is to throw a light on the various properties of bacterial cellulose so that a non-biology-based researchers or students can also understand the uniqueness of it. The chapter mainly focused on the textile and apparel application of bacterial cellulose. The first part details the physical, chemical, thermal properties of bacterial cellulose in detail along with the effect of in-situ and ex-situ modifications. Similarly, the later part of the chapter focuses on scouring, bleaching, colouration and functionalisation of dried bacterial cellulose by focusing on textile applications. The section also summarises the effect of these treatments on the basic properties of the bacterial cellulose.

Table 3.6 Various chemicals and properties achieved with bacterial cellulose functionalisation process

S. no.	Chemical used	Property changes	Application area	Reference
1.	ε-poly-L-lysine (ex situ)	• No changes in the thickness • Fibril structure visible with 200 nm average pore size • Increase in fibre diameter • No changes in modulus upon storage • Antimicrobial effect	Antimicrobial wound dressing	Fürsatz et al. [97]
2.	C60 nanoparticles (ex situ)	• No changes in three-dimensional fibrous network structure • Generates abundant reactive singlet oxygen • Potential antibacterial property. 95% antibacterial property noted in the light and in the absence 50% noted • Bacterial cellulose—C60 composite is an effective photosensitiser that can release abundant ROS to kill cancer cells in the presence of light • Death rate of the A-431 cells exceeded 80%, and the best anticancer effects were achieved	Antimicrobial wound dressing and for cancer therapy	Chu et al. [98]
3.	Chimeric protein containing a cellulose-binding module and an adhesion peptide (ex situ)	• Improvement in the adhesion of human microvascular endothelial cells • (HMEC) to bacterial cellulose noted a successful ingrowth of adhesion peptide treated bacterial cellulose noted with human cell seeding • Proteins containing a cellulose-binding module domain, with high affinity aids a special control of interaction mechanism with cellulose • Results suggested a potential use of bacterial cellulose along with proteins containing a cellulose-binding module domain in in vitro or in vivo applications	Blood vessel replacement	Andrade et al. [99]

(continued)

Table 3.6 (continued)

S. no.	Chemical used	Property changes	Application area	Reference
4.	Poly(fluorophenol) (ex situ)	• Increase in water contact angle from 54.5° (untreated) to 120° • Decrease in surface energy from 51.12 to 27.9 mN/m • Water absorption time increases to 5 min 45 s from 2 min 40 s • Increase in oil contact angle of 46.5° to 87.2° • The treatment increased the tensile strength threefold after laundry	Hydrophobicity and durability	Song et al. [100]
5.	6-Carboxyfluorescein-modified glucose (in situ)	• The modified cellulose developed green fluorescence based on UV and confocal laser scanning microscopy • Poor thermal stability noted than the native cellulose • Crystallinity percentage reduced • Higher pore size and numbers • Lower elastic modulus and tensile strength	Fluorescence on bacterial cellulose structure	Gao et al. [101]
6.	Hydroxyapatite and antibone morphogenetic protein 2 (ex situ)	• Wet cellulose material possessed better resistance to drilling value than the dry bacterial cellulose hydroxyapatite composite • Lower contact angle ensures the higher moisture property • Composite promoted higher cell viability • Non-cytotoxic, genotoxic and mutagenic biomaterial	As a biomaterial for bone filler material	Coelho et al. [102]

(continued)

Table 3.6 (continued)

S. no.	Chemical used	Property changes	Application area	Reference
7.	Silver nanoparticles (ex situ)	• Lower crystallinity percentage than native bacterial cellulose • Higher concentration of silver nanoparticle showed excellent bacterial resistance even after 100 h (OD value) • In disc diffusion method, maximum of 6.5 mm zone of inhibition noted for higher concentration • The antimicrobial activity increased with the silver nanoparticle concentration	Antibacterial membrane for wound-healing applications	Pal et al. [103]
8.	RGDC peptides (R: arginine; G: glycine; D: aspartic acid; C: cysteine) and gentamicin	• Composite material showed a lower porosity than the native bacterial cellulose • 3D network structure maintained • Thickness of the membrane increased • Higher antimicrobial activity by disc diffusion method against *Streptococcus mutans* • Promotes fibroblast adhesion and cell growth • Non-toxic	Wound-healing and drug delivery systems	Rouabhia et al. [104]
9.	Aminoalkylsilane	• Develops a lamellar structure with highly ordered fibres after being grafted • XPS analysis showed a higher amount of Si and N molecules on the surface • Increment in water contact angle noted than native bacterial cellulose • Reduction of *E. coli* attachment by 100%, *S. aureus* attachment by 99.4% and *B. subtilis* attachment by 99.9% • No cell cytotoxicity, however, lower cell growth noted than native cellulose	Antibacterial and antifungal activity	Shao et al. [8]

(continued)

Table 3.6 (continued)

S. no.	Chemical used	Property changes	Application area	Reference
10.	Phosphoric functional groups	• No effect found on physical structure and morphology • Structural aspect is noted as main reason for higher adsorption • The adsorption capacity noted good at pH 8 • With respect to time, more than 50% of the adsorption happened in first 20 min • Adsorption kinetics fixed well with Elovich model and pseudo-second-order model • Adsorption isotherms fitted with Langmuir model	U(VI) adsorbent	Zhuang and Wang [105]

References

1. Mohite BV, Patil SV (2014) Physical, structural, mechanical and thermal characterization of bacterial cellulose by G. hansenii NCIM 2529. Carbohydr Polym 106:132–141
2. Ul-Islam M, Khana T, Park JK (2012) Water holding and release properties of bacterial cellulose obtained by in situ and ex situ modification. Carbohyd Polym 88:596–603
3. Chen H-H, Chen L-C, Huang H-C, Lin S-B (2011) In situ modification of bacterial cellulose nanostructure by adding CMC during the growth of *Gluconacetobacter xylinus*. Cellulose 18:1573–1583
4. Khandelwal M, Windle AH, Hessler N (2016) In situ tunability of bacteria produced cellulose by additives in the culture media. J Mater Sci 51:4839–4844
5. Rukaa DR, Simon GP, Deana KM (2013) In situ modifications to bacterial cellulose with the water insoluble polymer poly-3-hydroxybutyrate. Carbohyd Polym 92:1717–1723
6. Teixeira SRZ, dos Reis EM, Apati GP, Meier MM, Nogueira AL, Garcia MCF, dos Santos Schneidera AL, Pezzin APT, Porto LM (2019) Biosynthesis and functionalization of bacterial cellulose membranes with cerium nitrate and silver nanoparticles. Mater Res 22(suppl. 1):e20190054
7. Pal S, Nisi R, Stoppa M, Licciulli A (2017) Silver-functionalized bacterial cellulose as antibacterial membrane for wound-healing applications. ACS Omega 2:3632–3639
8. Shao W, Wu J, Liu H, Ye S, Jiang L, Liu X (2017) Novel bioactive surface functionalization of bacterial cellulose membrane. Carbohyd Polym 178:270–276
9. Cazón P, Vázquez M, Velazquez G (2019) Composite films with UV-barrier properties of bacterial cellulose with glycerol and poly(vinyl alcohol): puncture properties, solubility, and swelling degree. Biomacromolecules. https://doi.org/10.1021/acs.biomac.9b00704
10. Rathinamoorthy R, Aarthi T, Aksaya Shree CA, Haridharani P, Shruthi V, Vaishnikka RL (2019) Development and characterization of self-assembled bacterial cellulose nonwoven film. J Nat Fibers. https://doi.org/10.1080/15440478.2019.1701609
11. Moharram MA, Mahmoud OM (2008) FTIR spectroscopic study of the effect of microwave heating on the transformation of cellulose I into cellulose II during mercerization. J Appl Polym Sci 107:30–36
12. Oh SY, Yoo DI, Shin Y, Kim HC, Kim HY, Chung YS, Park WH, Youk JH (2005) Crystalline structure analysis of cellulose treated with sodium hydroxide and carbon dioxide by means of X-ray diffraction and FTIR spectroscopy. Carbohyd Res 340:2376–2391
13. Wonga S-S, Kasapis S, Tan YM (2009) Bacterial and plant cellulose modification using ultrasound irradiation. Carbohyd Polym 77:280–287. https://doi.org/10.1016/j.carbpol.2008.12.038
14. Focher B, Palma MT, Canetti M, Torri G, Cosentino C, Gastaldi G (2001) Structural differences between non-wood plant celluloses: evidence from solid state NMR, vibrational spectroscopy and X-ray diffractometry. Ind Crops Prod 13:193–208
15. Movasaghi Z, Rehman S, Rehman I (2008) Fourier transform infrared (FTIR) spectroscopy of biological tissues. Appl Spectrosc Rev 43:134–179
16. Yassine F, Bassil N, Chokr A, Samrani AE, Serghei A, Boiteux G, Tahchi ME (2016) Two-step formation mechanism of *Acetobacter cellulosic* biofilm: synthesis of sparse and compact cellulose. Cellulose 23:1087–1100. https://doi.org/10.1007/s10570-016-0884-3
17. Kacurakova M, Smith AC, Gidley MJ, Wilson RH (2002) Molecular interactions in bacterial cellulose composites studied by 1D FT-IR and dynamic 2D FT-IR spectroscopy. Carbohyd Res 337:1145–1153
18. Han J, Shim E, Kim HR (2019) Effects of cultivation, washing, and bleaching conditions on bacterial cellulose fabric production. Text Res J 89(6):1094–1104
19. Song JE, Cavaco-Paulo A, Silva C (2020) Improvement of bacterial cellulose nonwoven fabrics by physical entrapment of lauryl gallate oligomers. Text Res J 90(2):166–178
20. Yu X, Atalla RH (1996) Production of cellulose II by *Acetobacter xylinum* in the presence of 2,6-dichlorobenzonitrile. Int J Biol Macromol 19:145–146

21. Maeda H, Nakajima M, Hagiwara T, Sawaguchi T, Yano S (2006) Bacterial, cellulose/silica hybrid fabricated by mimicking biocomposites. J Mater Sci 41(17):564–565
22. Kulkarni PK, Dixit SA, Singh UB (2012) Evaluation of baceterial cellolose produced from Acetobacter xylini, as Pharmaceutical excipient. Am J Drug Discovery Dev 2(2):72–86
23. Tsouko E, Kourmentza C, Ladakis D, Kopsahelis N, Mandala I, Papanikolaou S, Paloukis F, Alves V, Koutinas A (2015) Bacterial cellulose production from industrial waste and by-product streams. Int J Mol Sci 16:14832–14849
24. Rohaeti E, Endang Widjajanti Laksono FX, Rakhmawati A (2017) Characterization and the activity of bacterial cellulose prepared from rice waste water by addition with Glycerol and chitosan. ARPN J Agric Biol Sci 12(8):241–248
25. George J, Ramanab KV, Sabapathy SN, Jagannath JH, Bawa AS (2005) Characterization of chemically treated bacterial (*Acetobacter xylinum*), biopolymer: some thermo-mechanical properties. Int J Biol Macromol 37:189–194
26. Cheng K-C, Catchmark JM, Demirci A (2009) Enhanced production of bacterial cellulose by using a biofilm reactor and its material property analysis. J Biol Eng 3:12
27. Jia Y, Wang X, Huo M, Zhai X, Li F, Zhong C Preparation and characterization of a novel bacterial cellulose/chitosan bio-hydrogel. Nanomaterials Nanotechnol 7:1–8
28. Barud HS, Ribeiro CA, Crespi MS, Martines MAU, Dexpert-Ghys J, Marques RFC, Messaddeq Y, Ribeiro SJL (2007) Thermal characterization of bacterial cellulose–phosphate composite membranes. J Therm Anal Calorim 87(3):815–818
29. Barud HS, de Araújo Júnior AM, de Assunção RMN, Meireles CS, Cerqueira DA, Filho GR, Messaddeq Y, Ribeiro SJL (2007) Thermal characterization of cellulose acetate produced from homogeneous acetylation of bacterial cellulose. Thermochim Acta 471(1):61–69
30. Yim SM, Song JE, Kim HR (2017) Production and characterization of bacterial cellulose fabrics bynitrogen sources of tea and carbon sources of sugar. Process Biochem 59:26–36
31. Hassan EA, Abdelhady HM, Sohir S. El-Salam A, Abdullah SM (2015) The characterization of bacterial cellulose produced by *Acetobacter xylinum* and Komgataeibacter saccharovorans under optimized. Br Microbiol Res J 9(3):1–13
32. Gea S, Reynolds CT, Roohpour N, Wirjosentono B, Soykeabkaew N, Bilotti E, Peijs T (2011) Investigation into the structural, morphological, mechanical and thermal behaviour of bacterial cellulose after a two-step purification process. Biores Technol 102:9105–9110
33. Fernandes M, Gama M, Dourado F, Souto AP (2019) Development of novel bacterial cellulose composites for the textile and shoe industry. Microb Biotechnol 12(4):650–661
34. Gelin K, Bodin A, Gatenholm P, Mihranyan A, Edwards K, Stromme M (2007) Characterization of water in bacterial cellulose using dielectric spectroscopy and electron microscopy. Polymer 48:7623–7631
35. Cazón P, Velázquez G, Vázqueza M (2020) Bacterial cellulose films: evaluation of the water interaction. Food Packag Shelf Life 25:100526
36. Rebelo AR, Archer AJ, Chen X, Liu C, Yang G, Liu Y (2018) Dehydration of bacterial cellulose and the water content effects on its viscoelastic and electrochemical properties. Sci Technol Adv Mater 19(1):203–211
37. Schrecker ST, Gostomski PA (2005) Determining the water holding capacity of microbial cellulose. Biotech Lett 27:1435–1438
38. Lin W-C, Lien C-C, Yeh H-J, Yu C-M, Hsu S (2013) Bacterial cellulose and bacterial cellulose–chitosan membranes for wound dressing applications. Carbohydr Polym 94:603–611
39. Roy S, Gennadios A, Weller CL, Testin RF (2000) Water vapor transport parameters of a cast wheat gluten film. Ind Crops Prod 11:43–50
40. Phisalaphong M, Suwanmajo T, Tammarate P (2008) Synthesis and characterization of bacterial cellulose/alginate blend membranes. J Appl Polym Sci 107:3419–3424
41. Cazón P, Vázquez M, Velazquez G (2020) Environmentally friendly films combining bacterial cellulose, chitosan, and polyvinyl alcohol: effect of water activity on barrier, mechanical, and optical properties. Biomacromolecules 21:753–760
42. Tome LC, Brandao L, Mendes AM, Silvestre AJD, Neto CP, Gandini A, Freire CSR, Marrucho IM (2010) Preparation and characterization of bacterial cellulose membranes with tailored surface and barrier properties. Cellulose 17:1203–1211

43. Leal S, Cristelo C, Silvestre S, Fortunato E, Sousa A, Alves A, Correia DM, Lanceros-Mendez S, Gama M (2020) Hydrophobic modification of bacterial cellulose using oxygen plasma treatment and chemical vapor deposition. Cellulose. https://doi.org/10.1007/s10570-020-03005-z
44. Wu X (2018) Control of the mechanical behavior of bacterial cellulose by mercerization. Master of Science thesis, Case Western Reserve University
45. Borysiak S, Garbarczyk J Applying the WAXS method to estimate the supermolecular structure of cellulose fibres after mercerisation. Fibres Text Eastern Europe, 11(5 (44)):104–106
46. Okano T, Sarko A (1985) Mercerization of cellulose. 2. Alkali cellulose intermediates and a possible mercerization mechanism. J Appl Polym Sci 30(1):325–332
47. Gea S (2010) Innovative bio-nanocomposites based on bacterial cellulose. Doctor of Philosophy, thesis, Queen Mary University of London, London
48. Kamal ASM, Misnon MI, Fadil F (2020) The effect of sodium hydroxide concentration on yield and properties of bacterial cellulose membranes. In: The 1st annual technology, applied science and engineering conference, IOP conference series: materials science and engineering, vol 732, p 012064
49. Sutrisno TA, Suryanto H, Wulandari R, Muhajir M, Shahrul SM, Zahari NS (2019) The effect of chemical pretreatment process on mechanical properties and porosity of bacterial cellulose film. J Mech Eng Sci Technol 3(1):8–17
50. Meftahi A, Khajavi R, Rashidi A, Rahimi MK, Bahador A (2015) Effect of purification on nano microbial cellulose pellicle properties. Procedia Mater Sci 11:206–211
51. Al-Shamary EE, Al-Darwash AK (2013) Influence of fermentation condition and alkali treatment on the porosity and thickness of bacterial cellulose membranes. Online J Sci Technol 3(2)
52. Indrarti L, Yudianti R (2012) Morphological and thermal properties of alkali treated bacterial cellulose from coconut water. Indonesian J Mater Sci 13(3):221–222
53. Suryanto H, Sutrisno TA, Muhajir M, Zakia N, Yanuhar U (2018) Effect of peroxide treatment on the structure and transparency of bacterial cellulose film. In: MATEC web of conferences, vol 204, No 05015. https://doi.org/10.1051/matecconf/201820405015
54. Jahan F (2014) Bacterial cellulose: production, properties, scale up and industrial applications. Doctor of philosophy thesis, University of Delhi. http://hdl.handle.net/10603/26212
55. Gayathry G, Gopalaswamy G (2014) Production and characterisation of microbial cellulosic fibre from *Acetobacter xylinum*. Indian J Fibre Text Res 39:93–96
56. Solatorio N, Chong Liao C (2019) Synthesis of cellulose by *Acetobacter xylinum*: a comparison vegan leather to animal and imitation leather. Honors thesis, University of Wyoming, Spring
57. Fairs M (2014) Microbes are "the factories of the future". https://www.dezeen.com/2014/02/12/movie-biocouture-microbes-clothing-wearable-futures/
58. Ghalachyan A (2018) Evaluation of consumer perceptions and acceptance of sustainable fashion products made of bacterial cellulose. Graduate Theses and Dissertations, 16583. https://lib.dr.iastate.edu/etd/16583
59. Ng FMC, Wang PW (2016) Natural self-grown fashion from bacterial cellulose: a paradigm shift design approach in fashion creation. Des J 19:837–855
60. Zhong C (2010) Colored bacteria cellulose product and preparation method thereof, CN102127576A
61. Tyurin I, Getmantseva V, Andreeva E, Kashcheev O (2019) The study of the molding capabilities of bacterial Cellulose. In: AUTEX2019—19th world textile conference on textiles at the crossroads, 11–15 June 2019, Ghent, Belgium
62. Lv P, Lu X, Zhou H, Sun X Biosynthesis of bacterial cellulose for in-situ assembly of intelligent packaging with natural dyes—life-made intelligent packaging. BioResources 15(2):2111–2113
63. Wood D, Liu H, Salusso CJ (2015) Production and characterization of bacterial cellulose fabrics. In: International Textile and Apparel Association (ITAA) annual conference proceedings, vol 143. https://lib.dr.iastate.edu/itaa_proceedings/2015/posters/143

64. Shim E, Kim HR (2019) Coloration of bacterial cellulose using in situ and ex situ methods. Text Res J 89(7):1297–1310
65. Gao M, Li J, Bao Z, Hu M, Nian R, Feng D, An D, Li X, Xian M, Zhang H (2019) A natural in situ fabrication method of functional bacterial cellulose using a microorganism. Nat Commun 10:437
66. Harmon J (2017) Homegrown: investigating design potential of bacterial cellulose. In: International Textile and Apparel Association (ITAA) annual conference proceedings, vol 15. https://lib.dr.iastate.edu/itaa_proceedings/2017/design/15
67. Costa AFS, de Amorim JDP, Almeida FCG, de Lima ID, de Paiva SSC, Rocha MAV, Vinhas GM, Sarubbo LA (2018) Dyeing of bacterial cellulose films using plant based natural dyes. Biomac. https://doi.org/10.1016/j.ijbiomac.2018.10.066
68. Harmon J (2019) Passive dyeing bacterial cellulose: results of a contact and low heat method. In: ITAA proceedings, #76. https://www.iastatedigitalpress.com/itaa/article/8756/galley/8408/view/
69. Song JE, Su J, Noro J, Cavaco-Paulo A, Silva C, Kim HR (2018) Bio-coloration of bacterial cellulose assisted by immobilized laccase. AMB Expr 8:19
70. Miyamoto H, Tsuduki M, Ago M, Yamane C, Ueda M, Okajima K (2014) Influence of dyestuffs on the crystallinity of a bacterial cellulose and a regenerated cellulose. Text Res J 84(11):1147–1158
71. Moisture Content Determination https://www.mt.com/in/en/home/applications/Laboratory_weighing/moisture-content-determination.html#:~:text=What%20is%20the%20drying%20oven,drying%20and%20determining%20the%20difference. Accessed 25 June 2020
72. Sederavičiūtė F, Domskienė J, Baltina I (2019) Influence of drying temperature on tensile and bursting strength of bacterial cellulose biofilm. Mater Sci 25(3)
73. Domskiene J, Sederaviciute F, Simonaityte J (2019) Kombucha bacterial cellulose for sustainable fashion. Int J Clothing Sci Technol 31(5):644–652. https://doi.org/10.1108/IJCST-02-2019-0010
74. Ratti C (2001) Hot air and freeze-drying of high-value foods: a review. J Food Eng 49(4):311–319. https://doi.org/10.1016/s0260-8774(00)00228-4
75. Illa MP, Sharma CS, Khandelwal M (2019) Tuning the physiochemical properties of bacterial cellulose: effect of drying conditions. J Mater Sci 54:12024–12035
76. Choudhary A (2017) Principle of tray dryer. https://www.pharmaguideline.com/2017/09/principle-of-tray-dryer.html?m=1#gsc.tab=0. Accessed 25 June 2020
77. Pa'e N, Hamid NIA, Khairuddin N, Zahan KA, Seng KF, Siddique BM, Muhamad II (2014) Effect of different drying methods on the morphology, crystallinity, swelling ability and tensile properties of *Nata De Coco*. Sains Malays 43(5):767–773
78. Vasconcellos V, Farinas C (2018) The effect of the drying process on the properties of bacterial cellulose films from gluconacetobacter hansenii. Chem Eng Trans 64:145–150. https://doi.org/10.3303/CET1864025
79. Zeng M, Laromaine A, Roig A (2014) Bacterial cellulose films: influence of bacterial strain and drying route on film properties. Cellulose 21:4455–4469
80. Jennifer H, Logan F, Natalie T (2020) Exploring the potential of bacterial cellulose for use in apparel. J Text Sci Fashion Tech 5(2). JTSFT.MS.ID.000606. https://doi.org/10.33552/jtsft.2020.05.000606
81. Tsotsas E, Mujumdar AS (2011) Modern drying technology, vol 3. In: Product quality and formulation. John Wiley & Sons, Hoboken, pp 185
82. Every H (2017) Drying with supercritical Co_2. https://www.tasteofscience.com/articles/1275/drying-with-supercritical-co2.html#:~:text=When%20the%20processing%20temperature%20and,is%20in%20a%20supercritical%20state.&text=This%20allows%20the%20CO2%20to,hours%2C%20and%20reduced%20energy%20consumption. Accessed 25 June 2020
83. Antal T (2015) Comparative study of three drying methods: freeze, hot air assisted freeze and infrared-assisted freeze modes. Agron Res 13(4):863–878

84. Zhang C, Wang L, Zhao J, Zhu P (2011) Effect of drying methods on structure and mechanical properties of bacterial cellulose films. Adv Mater Res 239–242:2667–2670. https://doi.org/10.4028/www.scientific.net/AMR.239-242.2667
85. Vacuum drying https://en.wikipedia.org/wiki/Vacuum_drying. Accessed 25 June 2020
86. Orsat V, Changrue V, Raghavan VGS (2006) Microwave drying of fruits and vegetables. Stewart Postharvest Rev 6:4
87. Indriyati I, Irmawati Y, Puspitasari T (2019) Comparative study of bacterial cellulose film dried using microwave and air convection heating. J Eng Technol Sci 51(1):121–132
88. Azadimanesh F, Mohammadi N (2015) A plasticizer index to universally correlate the normalized work of fracture and elastic modulus of plasticized cellulose triacetates. Carbohyd Polym 130:316–324
89. Sun Y et al (2018) The effects of two biocompatible plasticizers on the performance of dry bacterial cellulose membrane: a comparative study. Cellulose 25(10):5893–5908
90. Cielecka I et al (2019) Glycerol-plasticized bacterial nanocellulose-based composites with enhanced flexibility and liquid sorption capacity. Cellulose 26(9):5409–5426
91. Cazón P, Vázquez M, Velazquez G (2019) Composite films with UV-barrier properties of bacterial cellulose with glycerol and poly (vinyl alcohol): puncture properties, solubility, and swelling degree. Biomacromolecules 20(8):3115–3125
92. Indrarti L et al (2016) Physical and mechanical properties of modified bacterial cellulose composite films. In: AIP conference proceedings, AIP Publishing LLC
93. Rohaeti E, Laksono E, Rakhmawati A (2017) Characterization and the activity of bacterial cellulose prepared from rice waste water by addition with glycerol and chitosan. Agric Biol Sci 12(8):241–248
94. Faridah F et al (2014) Effect of plasticizer and fermetation time on cellulose membrane production and analysis of material property. In: 4th Syiah Kuala University annual international conference 2014, Syiah Kuala University
95. Meftahi A et al (2018) Preventing the collapse of 3D bacterial cellulose network via citric acid. J Nanostruct Chem 8(3):311–320
96. Teixeira SRZ et al (2019) Biosynthesis and functionalization of bacterial cellulose membranes with cerium nitrate and silver nanoparticles. Mater Res 22
97. Fürsatz M et al (2018) Functionalization of bacterial cellulose wound dressings with the antimicrobial peptide ε-poly-L-Lysine. Biomed Mater 13(2):025014
98. Chu M et al (2018) Functionalization of composite bacterial cellulose with C 60 nanoparticles for wound dressing and cancer therapy. RSC Adv 8(33):18197–18203
99. Andrade FK et al (2010) Improving bacterial cellulose for blood vessel replacement: functionalization with a chimeric protein containing a cellulose-binding module and an adhesion peptide. Acta Biomater 6(10):4034–4041
100. Song JE et al (2019) Functionalization of bacterial cellulose nonwoven by poly (fluorophenol) to improve its hydrophobicity and durability. Front Bioeng Biotechnol 7:332
101. Gao M et al (2019) A natural in situ fabrication method of functional bacterial cellulose using a microorganism. Nat Commun 10(1):1–10
102. Coelho F et al (2019) Bacterial cellulose membrane functionalized with hydroxiapatite and anti-bone morphogenetic protein 2: a promising material for bone regeneration. PLoS ONE 14(8):e0221286
103. Pal S et al (2017) Silver-functionalized bacterial cellulose as antibacterial membrane for wound-healing applications. ACS Omega 2(7):3632–3639
104. Rouabhia M et al (2014) Production of biocompatible and antimicrobial bacterial cellulose polymers functionalized by RGDC grafting groups and gentamicin. ACS Appl Mater Interfaces 6(3):1439–1446
105. Zhuang S, Wang J (2019) Removal of U (VI) from aqueous solution using phosphate functionalized bacterial cellulose as efficient adsorbent. Radiochim Acta 107(6):459–467

Chapter 4
Applications of Bacterial Cellulose

Focus on Clothing and Accessories

4.1 Introduction

Bacterial cellulose is a very strong and potential biopolymer that can be applied in various industrial sectors such as textiles and fashion, health care and medicine, food and agriculture. Bacterial cellulose has a uniqueness which is its chemical purity and contrary to cellulose from plants which are usually associated with hemicellulose and lignin, removal of them is highly complicated. The bounty of bacterial cellulose is its biocompatibility, biodegradability, ultra-fine molecular structure, desired mechanical strength, non-toxic nature, purity and neutrality. It comprises nanofibrillar structure which is responsible for creating a microporous structure along with a large specific surface area which is the main reason for creating enormous opportunities for its modification, and hence it is highly useful to produce various types of composite materials [1–4]. The chapter details various applications of bacterial cellulose and its complex forms. The primary focus is given on the textile and clothing applications of the bacterial cellulose, where the various research works summarised to enlighten the development in the fashion sector.

Next to the clothing application, the potent as a leather substitution material is detailed. The possible applications like shoes, jackets, handbags and clutch as an alternative sustainable material were enlisted with the research details. The review will allow the readers to understand the application possibilities as fashion accessories over the clothing. Several research works showed their unique issues and possible solutions to commercialise the product in the market. Later part of the chapter details the medical and healthcare applications followed by engineering applications of bacterial cellulose.

© The Author(s), under exclusive license to Springer Nature Singapore Pte Ltd. 2021

S. S. Muthu and R. Rathinamoorthy, *Bacterial Cellulose*, Sustainable Textiles: Production, Processing, Manufacturing & Chemistry,
https://doi.org/10.1007/978-981-15-9581-3_4

4.2 Clothing Application of Bacterial Cellulose

In the textile and fashion industry, bacterial cellulose produced from the static culture method is used mainly due to the structural formation. The three-dimensional nanofibrous non-woven mat structure is mainly obtained after a predefined period of cultivation. The cultivated material is immediately purified and used for subsequent application. Concerning the clothing applications, dried bacterial cellulose material is commonly used in contrast to other applications. It is also necessary to mention here while considering the clothing application that most of the researchers used kombucha-based bacterial cellulose for clothing production. The preliminary work on fashion application of bacterial cellulose was identified and performed by designer Suzanne Lee. She is a research fellow in the school of fashion and textiles at Central Saint Martins in London in the year 2003. In her research, she coined the name 'biocouture', which represents the production of fabric naturally from sustainable sources. This term biocouture broadly covers not only the fabric produced from bacteria or microorganisms but also other sustainable materials like chitin, mushrooms, natural cellulose, etc. In her research, she developed cellulose fabric using *Gluconacetobacter xylinus*. The bacteria ferment the tea sugar media and developed a cellulose mat on the surface of the liquid. In the bulk production, by immersing the 3D dress form into a large amount of fermentation medium and so it can allow the bacteria to grow around the mould. The researcher also mentioned that depending on the post-processing techniques and drying method, the product will be durable than a normal low cost knitted t-shirt [5]. Later in 2012, she started a design consultancy with the name 'BioCouture' which advises brands to produce sustainable products. She has partnered with various leading brands. Currently, she is associated with American bioleather company Modern Meadow [6]. The developed bacterial cellulose after drying becomes a skin-like material with optimal stiffness which can be moulded or sewed into shirts or jackets. The researchers also added vegetables like beetroot and blueberries to colour them and she ended up with new designs, colours and patterns. Figure 4.1 represents a garment developed by designer Lee [7].

Though S. Lee developed apparel using bacterial cellulose, after her biocouture project she started working on several other materials that are sustainable. No technical studies performed on the bacterial cellulose properties and issues related to enduses. Based on her preliminary work, several research works performed after 2012 in the technical aspects. Ng and Wang evaluated the comfort-oriented perceptions of the developed bacterial cellulose fabric. In their research, they developed bacterial cellulose fabric using milk, red wine, green tea, red tea and coconut milk. In the analysis of developed fabric, the fabric produced from sources like milk, wine and coconut milk is lower in tensile strength and poor handling properties. Out of the selected medium, both the tea sources yielded higher bacterial cellulose.

However, the green tea can able to produce more amount of pellicles in a shorter time than the red tea. Hence, the researcher developed bacterial cellulose using green tea with different concentrations of 5 and 15 g/L. The cultivation time varied as 4, 6

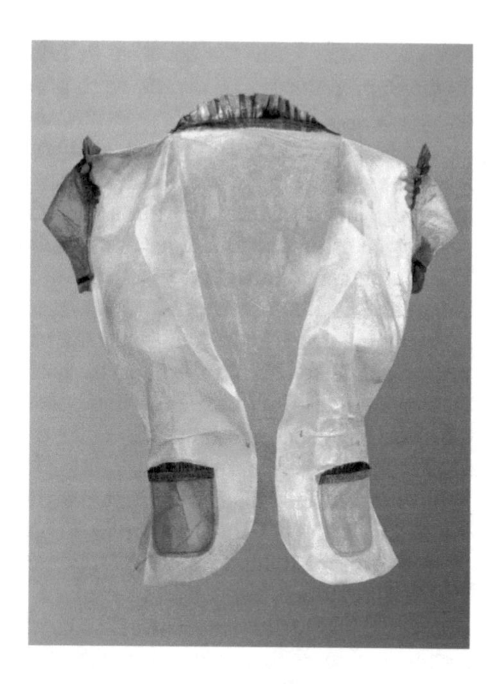

Fig. 4.1 Garment developed by designer Suzanne Lee as a part of her project [7]

and 8 days. The developed bacterial cellulose fabric was analysed for comfort properties using one hundred and fifty participants in terms of hand comfort, flexibility comfort and breathability comfort. The subjects were provided with a 20 cm × 20 cm sample to rate the property on a 5-point Likert scale. In the case of hand comfort and flexibility comfort, a maximum rating of 4.27 and 4.73 provided for 6 days grown 15 g/L tea used fabric. The breathability comfort is noted for 4 days grown lower tea gram used sample due to their lower thickness. In the case of colour and texture values, all the samples were rated as 3 with a small deviation. Based on the overall comfort ranking, the 15 g/L tea used bacterial cellulose grown for 6 days obtained the first rank followed by 5 g/l tea sample with 8-day cultivation and 6-day cultivation in second and third. The research optimised the tea concentration and cultivation time in terms of tactile comfort properties of the bacterial cellulose [8].

In further research, they developed a true size three-dimensional seamless apparel with bacterial cellulose. The researcher proposed various theoretical concepts in their subsequent research for the development of bacterial cellulose-based apparels for real. The first method they represented was the conversion of 2D material into 3D material. This can be achieved by several other combined methods like cut and sewing of the 2D sheets or by growing material on small scaffolds and then sewing them together. Alternatively, matured bacterial cellulose can also be ground and applied as a paste on the 3D form to obtain the shape. The second method details the development of the direct formation of a three-dimensional structure by dipping a 3D dress form directly into the cultivation bath either by parts or as a whole. After the growth adhesion, the dummy can be rotated or moved for the next place where the formation is required. The third method was all about the inclusion of different foreign materials into the bacterial cellulose fabric to enhance the functional and aesthetic properties. The details were provided in Fig. 4.2 [9]. The 3D prototype developed by Ng and Wang is presented in Fig. 4.3 [9].

The higher water absorption capacity of the bacterial cellulose even after drying is noted as one of the major disadvantages in fashion applications. Hence, to impart the hydrophobic nature, the researcher treated the developed bacterial cellulose non-woven with two different concentrations of softener and hydrophobic agents. The researcher followed two methods to incorporate the hydrophobic nature into the bacterial cellulose by treating softener first, hydrophobic agent second, and hydrophobic agent first followed by the softener. The researchers performed water contact angle, crystallinity and surface morphological analysis. The results revealed that the treatment increased the water contact angle from 43° to 118°. This confirms the changes in the hydrophobic nature of the bacterial cellulose after treatment. Concerning the structural changes, the XRD analysis peaks revealed that the peak intensity of the treated cellulose noted higher than the control sample by confirming higher crystallinity. Similarly, the hydrophobic finished bacterial cellulose did not show any fibril structure over control fabric due to the coating of hydrophobic agent [10]. The durability properties of the bacterial cellulose are evaluated using tensile strength and abrasion resistance. The researcher developed bacterial cellulose and post-treated to remove impurities with 1% NaOH. The purified samples were glycerol treated and dried with air drying and freeze-drying method. The results were obtained to imply the application of bacterial cellulose in textile and fashion applications. The results reported that among the developed non-woven bacterial cellulose,

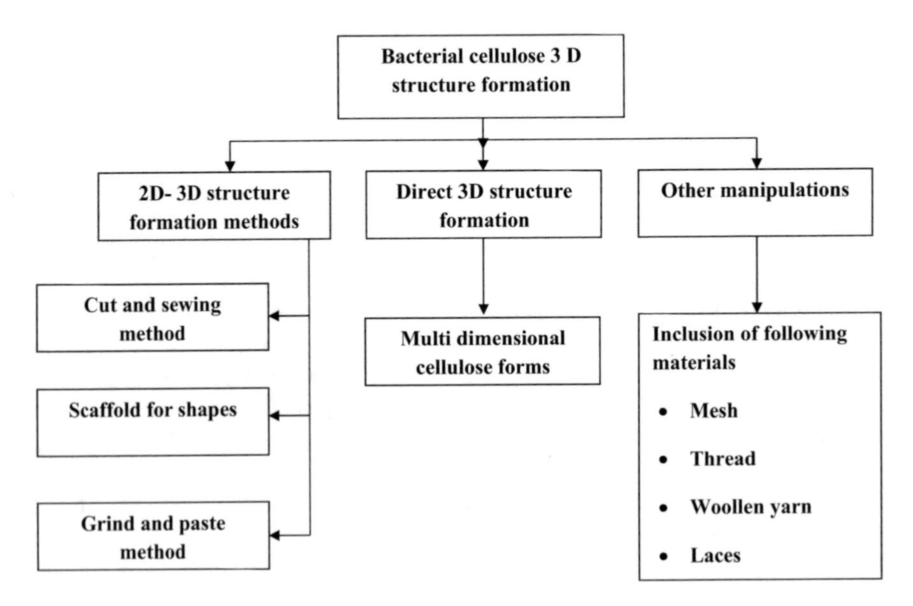

Fig. 4.2 The different potential methods to develop three-dimensional apparel using bacterial cellulose [9] [Authors own illustration]

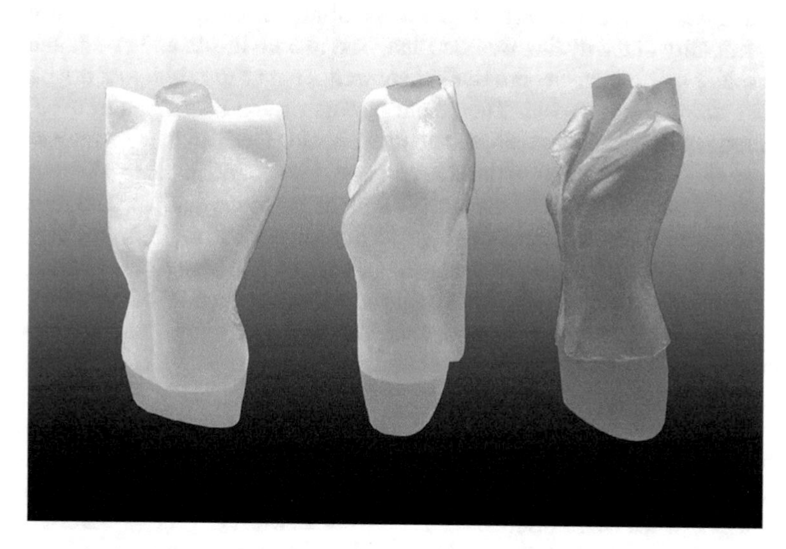

Fig. 4.3 Dry and wet prototypes of bacterial cellulose-based seamless 3D prototypes by Ng and Wang [9] [Reprinted with Permission]

higher tensile strength noted for the non-glycerol treated samples and dried with the freeze-drying method (61.42 N). Similarly, the non-glycerol treated samples from air, and freeze-dried method were stable enough to withstand 5000 cycles in Martindale tester with 9 kPa [11]. The author also evaluated the aesthetic and comfort properties of the selected samples. The hand values of the samples were analysed using AATCC-EP5. The results were promising in the case of glycerol treated freeze-dried fabric. They had shown higher pliability of 6.7 (mean) on a 7-point scale, but at the same time, the air-dried samples showed a poor average rating of 1.5. The freeze-dried sample also showed a higher rating for the softness ($M = 6$), whereas the air-dried sample is stiffer. Out of all the samples, the air-dried glycerol treated samples showed higher uniformity and lower thickness. However, the water absorbency results showed freeze-dried samples without glycerol treatment showed better absorbency [12].

Recently, few researchers analysed the technical side of the bacterial cellulose post-treatment with a detailed methodology. The researcher developed bacterial cellulose from HS medium with different sugars. They evaluated the effect of alkali treatment, neutralising treatment and bleaching process on various parameters like crystallinity, chemical structure, surface morphology and colour value of the product. The findings revealed that the NaOH treatment removed the impurities without altering the structure. Similarly, the bleaching process increased the crystallinity of the material, and the whiteness index 17 times higher than the produced bacterial cellulose. They have confirmed the three-step purification method proposed by the researcher effectively removes impurities like proteins and amino acids from the structure. Further, the purification process improved the internal fibril alignment, reduced crystal size, and showed higher crystallinity of the bacterial cellulose [13]. The potentiality of the dyeing was also analysed in both in-situ and ex-situ methods. The results reported that in-situ dyeing showed a better dyeability of the bacterial cellulose than ex-situ dyeing. The colour strength value of the in-situ dyed fabric also noted two times higher than ex-situ dyed fabric. The results were favour on the in-situ method as it provides better dye and fibre interaction in the production itself. This was reflected in terms of the surface roughness of the sample. The researcher proposed the in situ as a better technique, as this reduces the multiple steps involved in the ex-situ dyeing. Along with that the in-situ method also offers higher colour value than the other method [14]. Similarly, natural dyeing of bacterial cellulose also performed using natural components like turmeric, saffron and beetroot. The results of the research were supportive and showed a great potential for the bacterial cellulose [15].

Tyurin et al. analysed the mouldability of the bacterial cellulose for the apparel end use. In their study, they developed in-situ coloured bacterial cellulose after 2-week cultivation. After that, the impurities were removed using 2% NaOH treatment and dried for 2 days. The semi-dried material was moulded on body forms and dried using different drying methods. In study, they developed a prototype of a seamless mould for apparel end use as shown in Fig. 4.4 [16]. Similarly, other researchers developed vegetable-dyed and glycerine treated bacterial cellulose for apparel application. The design was drafted and freeze-dried samples were steamed again and moulded on

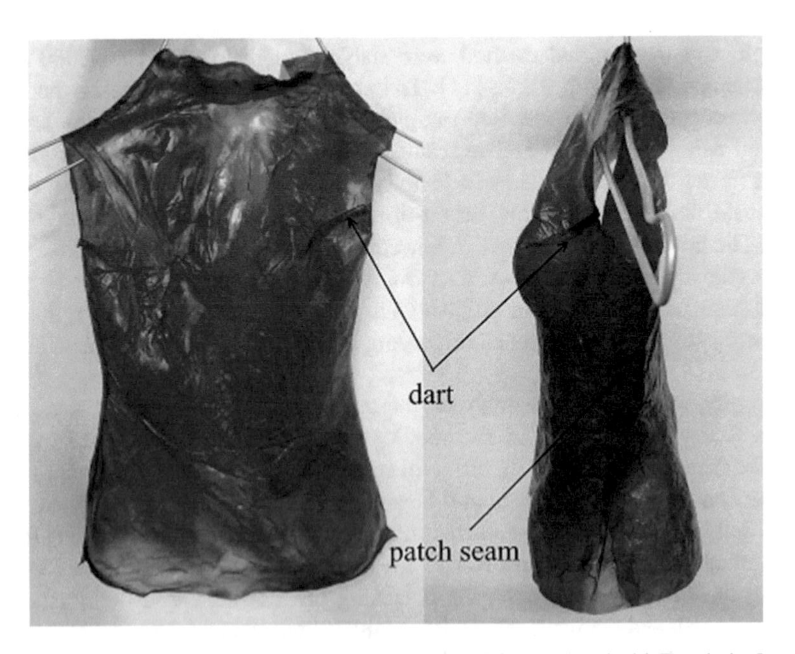

Fig. 4.4 Moulded garment form dyed bacterial cellulose [16] [Reprinted with Permission]

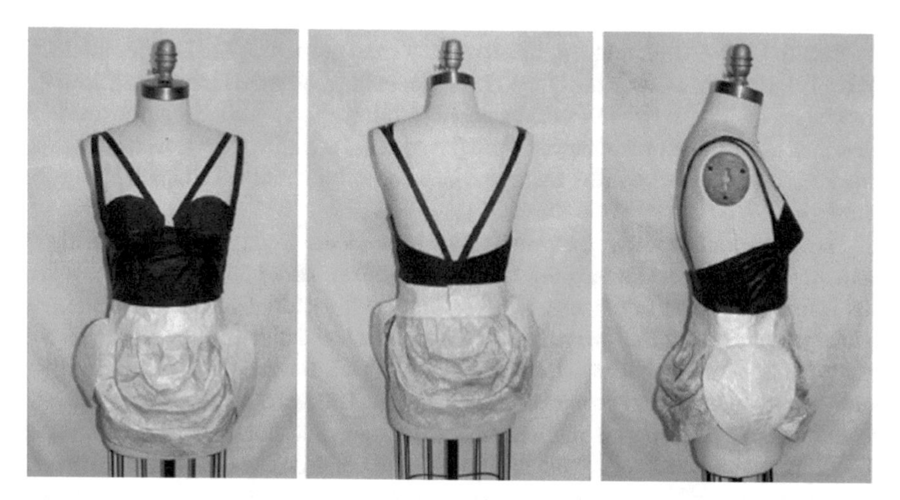

Fig. 4.5 Bacterial cellulose-based garment top and skirt [17] [Reprinted with permission]

the body forms. The material was again dried and stitched for apparel end use. They have reported the material imitated the leather and should not able to hold closures like zippers for application. Hence, several velcros are used in the construction to minimise the seams and stitches. The un-dyed material was used for the skirt. The designed top and skirt are provided in Fig. 4.5 [17].

Chan et al. developed a new methodology to develop tailor-shaped bacterial cellulose panels for textile product development. In their research, they used specific pattern-shaped containers to grow the bacterial cellulose. For example, the researchers used containers in the shape of the men's shirts sleeve, collar, yoke and bodices. The bacterial cellulose is cultivated on the containers for a long duration. The developed cellulose fabric on the specific pattern shape was given lining with the conventional lining fabric. The researcher proposed that using hot steam before sewing softens the fabric and provides better sewing ability. Similarly, they also reported the application of oil or wax makes the bacterial cellulose more flexible. The aesthetic look with wrinkle texture makes the bacterial cellulose fabric to look very similar to thin sheepskin leather. The researchers mentioned the method as one of the sustainable ways of developing fabric and tailor-shaped pattern development offers a zero-waste method for bulk production [18]. Concerning the drying properties, researchers reported that though bacterial cellulose material had impressive mechanical and technical properties it is lagging in its structural deformation properties. The use of wet bacterial cellulose for moulding the shape of the body provides a better elastic behaviour. Drying makes the bacterial cellulose to lose its strength and porous structure. Lower temperature drying around 25 °C preserves the properties. The experimental results confirmed the short term use of the product. The ageing properties of the bacterial cellulose showed that the properties of the fabric will change significantly during the storage time. The researcher recommended to store the material in the lower temperature to retain its best properties like tensile and thickness [19]. Recent research developed bacterial cellulose-based hydrogel by treating the glycerol with kombucha-based bacterial cellulose. The hydrogel also coated with stearic acid to protect the material from the external weather. The researcher developed wrist bands and t-shirts using the developed bacterial cellulose hydrogel as shown in Fig. 4.6.

The wear trial reported that after two weeks of usage the wrist band made of bacterial cellulose lost its flexibility. In the case of raw bacterial cellulose, the material tearing noted but in the case of glycerol treated bacterial cellulose the strength retained as such after the wear trial. Further, the material did not show any regrowth of the bacteria or any bacteria from human skin. This assured the clear sterilization and lower contamination property of the material. The researchers also conducted a feedback survey among the participants of the study and noted that the comfort of the wearing as neutral/good with an average rating of 4.3. The odour of the material noted as neutral with a rating of 3.3 in a 6-point scale. However, few participants reported smell due to the residual acid in the fabric. Neutral rating of 3 noted with respect to the visual appearance and the same rating of neutral (rating of 3) noted for the flexibility and softness of the material. Similarly, the participants also reported a satisfactory sweat absorption with the wrist band and no skin irritation or decolouration or itching noted during the wear study. The changes in the physical structure of the bacterial cellulose wrist band concerning the time are provided in Fig. 4.7 [20].

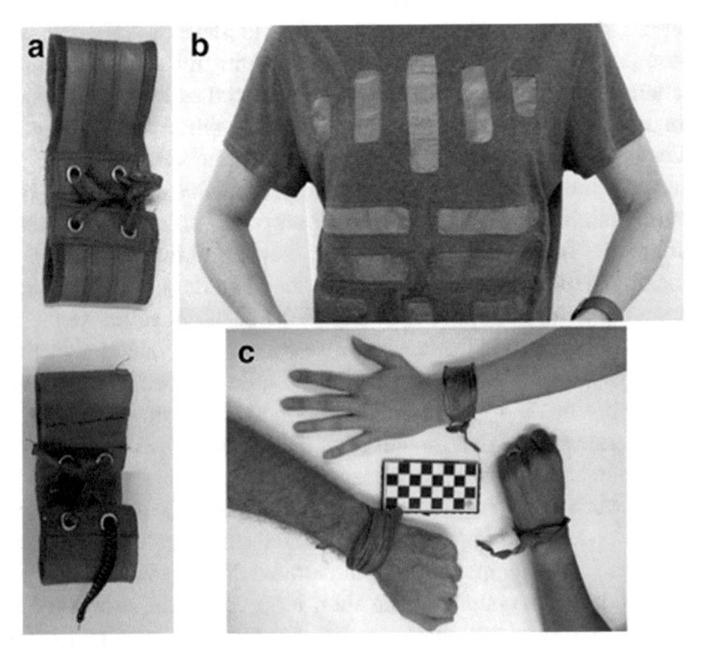

Fig. 4.6 Wrist band and t-shirt with bacterial cellulose fabric [20] [Reprinted with Permission]

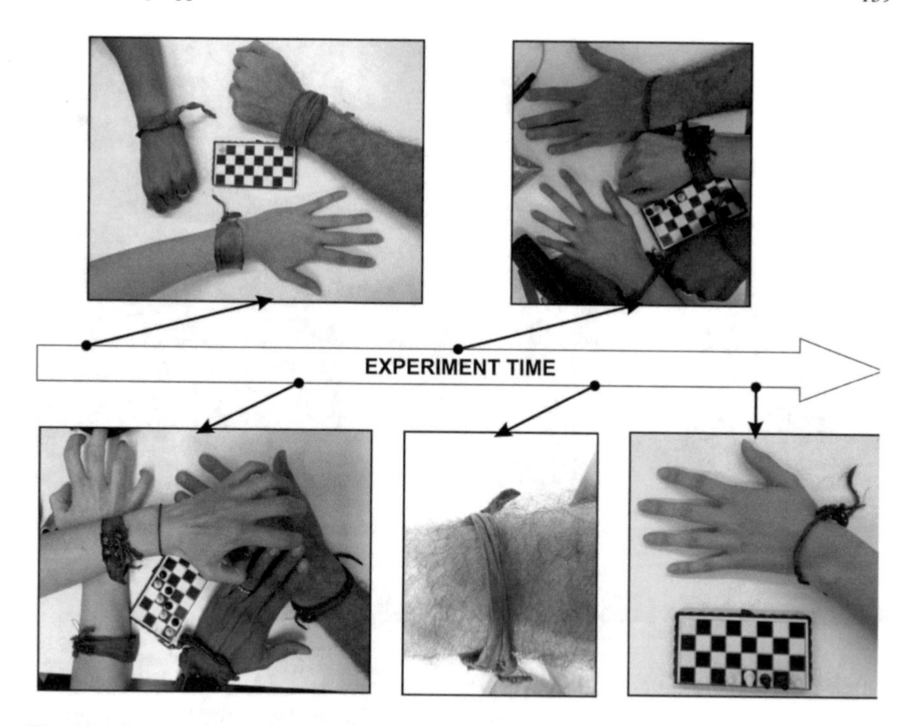

Fig. 4.7 Changes in the bacterial cellulose structure during the wear trial period of 2 weeks [20] [Reprinted with Permission]

A most recent research article summarised the various potential applications of bacterial cellulose as sustainable fashion material. It reported that the potentiality of the bacterial cellulose obtained from the kombucha tea. The research provided insight into new design possibilities of strength and flexibility of the material. The developed prototype is provided in Fig. 4.8. The possibilities of combining the bacterial cellulose with the origami techniques are also tried to create bio-based e-textiles. The loss of flexibility in the native bacterial cellulose fabric is one of the major disadvantages in the textile and fashion application. In order to overcome the issue, researchers attempted to incorporate the origami folds in the bacterial cellulose fabric. Additionally, the researcher incorporated the SMD-LED inside the structure of the bacterial cellulose using the self-adhesiveness of the fabric as an aid. The cultivated bacterial cellulose is primarily dyed with different colour and then sandwiched between the patterns of folds and dried completely to set the origami folds in the bacterial cellulose fabric. During the wet state itself, the conductive threads and SMD-LEDs were adhered to the fabric surface and dried. The resultant products called as 'AVA' by the researchers produced electrically active bio-based e-textile for the applications [22].

Fig. 4.8 'Biofabric-e' by Galina Mihaleva and Abhik Chowdhury [21] [Reprinted with Permission]

4.3 Bacterial Cellulose as Leather Alternative

Other than the clothing application, due to its thickness, look and feel, bacterial cellulose is commonly used as a leather alternative material for several applications. Particularly, the tactile properties of the bacterial cellulose imitate the finely finished leather, and further, it is also softer and more stretchable. The researchers also proposed several other advantages over animal skin and can be produced in the required shape, thickness, without defects and without any variations.

4.3.1 Fashion Accessories (Shoes and Bags)

Kombucha-based SCOBY is the commercially available form of the bacterial cellulose. Most of the apparel and leather alternative process used the kombucha as the source of bacterial cellulose. In this method, mostly tea is used as a nitrogen source and commercial sugar or glucose as a carbon source. Researchers used this medium for bacterial cellulose development and characterised the technical properties for various applications. The results reported that higher mechanical properties and surface appearance similar to the top grain leather. The researchers mentioned that for the same thickness, the tensile strength of the bacterial cellulose is two times higher than the top grain leather. However, the behaviour of low thickness is noted as one of the major disadvantages of bacterial cellulose after drying. The researcher did not perform any sewability and wearability analysis with the developed bacterial cellulose non-woven fabric in the applications. The developed bacterial cellulose layer and top grain leather compared by the researchers are provided in Fig. 4.9 [23].

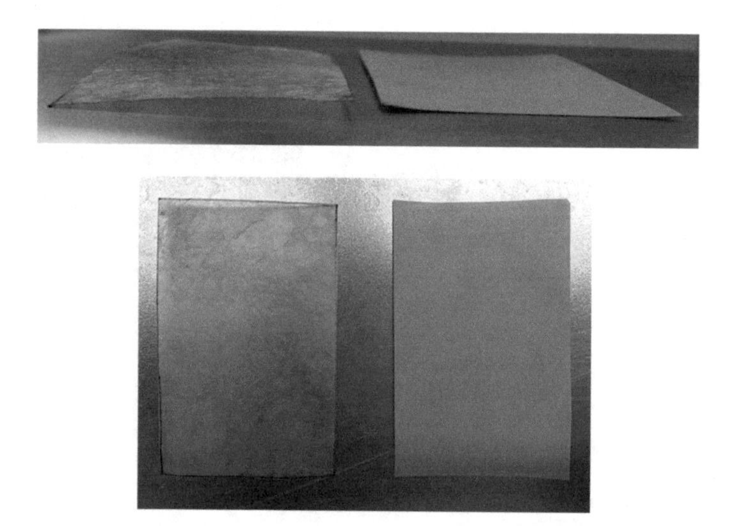

Fig. 4.9 Bacterial cellulose and top grain leather comparison [23] [Reprinted with Permission]

Using bacterial cellulose produced from sugar and tea using the kombucha medium, researchers developed leather alternative material. The researcher produced bacterial cellulose and evaluated the characteristics like the smell, feel, texture, overall liking of the material, etc. using subjective analysis. See-through structure, tanned colour, wrinkled texture and leather-like look are the common characteristics denoted by the participant on the positive side. Similarly, on the negative side, the participant reported an unpleasant smell on the material. This is mainly attributed to the cultivation medium of the kombucha culture. As the researcher did not perform any purification, the odour may be present in the fabric. Further, the participants were also questioned about the usage of bacterial cellulose as apparel through various attributes. The results indicated that the material cannot be used for clothing applications due to its see-through structure, rough texture and brittle nature. In the meantime, participants accepted the material as accessories like bags, hats and shoes [24]. Based on the results of the survey, researchers developed handbags and accessories like clutches using bacterial cellulose. Though several technical applications explored in the case of bacterial cellulose, this is observed as an important application as a leather alternative. With the laser-cut design in the front flap, a bacterial cellulose-based saddlebag was developed as illustrated in Fig. 4.10 [25].

Similarly, bacterial cellulose was suggested for the footwear application by several researchers. The main reason behind the proposal is due to the toxic nature of the leather product at the manufacturing, storage and also in the user phase. The use of chromium in footwear production process is one of the most toxic chemicals used. Tanning is known for its higher pollution among the various operation used in leather product manufacturing. Hence, when compared to the normal leather production, the production of bacterial cellulose is highly environmental friendly and no toxic chemicals are used. Along with that, the leather-like look and colour of the bacterial

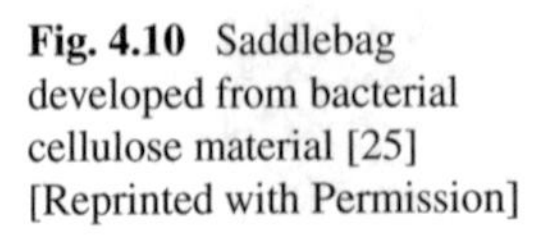

Fig. 4.10 Saddlebag developed from bacterial cellulose material [25] [Reprinted with Permission]

cellulose are one of the major reasons for its leather alternative applications. Due to the higher crystallinity and Young's modulus, the bacterial cellulose material showed its potential as a leather alternative. However, the researchers found that loss of structural flexibility upon drying is one of the major disadvantages in the end use as mentioned earlier. The higher rehydration rate after drying is another major disadvantage for apparel or clothing or leather application. Hence, to overcome this issue, the researcher used a plasticiser to increase the flexibility and a hydrophobic agent to reduce the hydrophilic nature of the bacterial cellulose. The characterisation results confirmed the hydrophobic nature and higher flexibility compared to the native or untreated bacterial cellulose. Based on the properties noted, the softener—hydrophobic agent added bacterial cellulose composite was suggested as a suitable material for leather alternative. The product was identified as a suitable material for using at the upper part of the shoe and other textile applications [26].

In another study, by inspiring the concept of Suzanne Lee's biocouture, researchers developed a retro shoe based on the concept of Scarlett and Rhett, leading characters from Gone with the Wind, one of the most famous American fiction bestsellers published in the 1930s. In this, kombucha-based SCOBY is used for bacterial cellulose production. To achieve the red colour, red onion skins were added to the cultivation. By using the dried bacterial cellulose, they developed bioleather shoes which are naturally self-coloured [27]. Other than kombucha material, the coconut milk-based bacterial cellulose is commonly known as Nata de coco also majorly used in the footwear application. The first work on Nata de coco was performed by Dr. Anselmo S. Cabigan and Arambulo of St. Paul University, QC, Philippines. They are the pioneers in the field and developed a shoe from the Nata-based bacterial cellulose. They made the bacterial cellulose as per the requirement of shoes and their products even passed the strength and durability test over the animal and artificial leather products [28]. Due to the higher increment in the number of diabetic patients around the world, few types of research suggested bacterial cellulose as a perfect material for diabetic ulcer shoes. By functionalising the bacterial cellulose or by developing

Fig. 4.11 **a** Footwear and **b** bags produced from the coconut water-based bacterial cellulose and natural fibre composite [30]

composites of it, the material made of bacterial cellulose may be a suitable alternative for diabetic patients. They have mentioned that properly monitored bacterial cellulose with the controlled condition along with medicine functionalisation will help to develop a perfect bacterial cellulose-based footwears and shoes for diabetic patients [29].

Recently, an Indian start-up company commercially producing bacterial cellulose-based footwear using coconut water instead of tea (Nata de coco). The founder of the company implemented the fermentation process as like kombucha but with coconut water. Their products were named as Malai. However, after the bacterial cellulose production, the manufacturing process also includes natural fibres like banana, hemp, etc. to obtain various aesthetic looks and textures. But they claim there is no synthetic or toxic material used in the production of the bacterial cellulose composite 'Malai'. They produced bags, shoes out of developed composites and are environment-friendly and biodegradable. Figure 4.11 represents the footwear developed from coconut water-based bacterial cellulose [30].

Alice Payne reported the potentiality of bacterial cellulose in the footwear manufacturing process. In her research, she produced kombucha-based bacterial cellulose and used it for sustainable footwear manufacturing. The hydrophilicity of the bacterial cellulose was one of the major issues. Hence, Alice used wax to provide a durable hydrophobic finish without affecting the tensile properties and comfortability of the material. They also outlined the use of zero-waste method in the shoes and footwear production from bacterial cellulose material [31].

4.3.2 Home Textile Materials

Out of clothing and textile applications, bacterial cellulose is also used in the home textile application. In the home textile application, Nikka Solatorio and Carolina

Chong Liao analysed the application of bacterial cellulose due to the negative impact of the leather products. In this research, researchers produced bacterial cellulose using HS medium and post-treated with 1% NaOH. The purified bleached bacterial cellulose is then treated with glycerol to improve the flexibility and allowed it to dry at room temperature. As an alternative to the leather ottomans, the bacterial cellulose-based material is developed for the ottoman application. They compared and analysed the properties of bacterial cellulose with animal leather and imitation leather ottomans. In results, they have reported that the developed bacterial cellulose ottoman whiteness index is reported less than the other leathers. Among the selected materials, original leather noted to be whiter and smooth in surface with a typical leather smell. With respect to the technical part, the sewability of the bacterial cellulose leathers is very easier than the imitation and animal leathers. It showed a paper-like look with a lot of surface wrinkles and with no leather-like smell on it. Out of all the samples, the animal leather noted as whitest and smoothest one. The imitation leather occupies the second place [32]. Figure 4.12 shows the bacterial cellulose bases ottoman developed by the researchers.

The researcher also reported that the cost of production of the bacterial cellulose is a bit higher than the imitation leather but lesser than the original leather. However, the production process of imitation and animal leather causes severe environmental pollution in terms of water, air and soil, whereas the bacterial cellulose production is environmental friendly since there is no cause of pollution in any form. The researcher also proposed to develop various home textile and clothing materials in future [32].

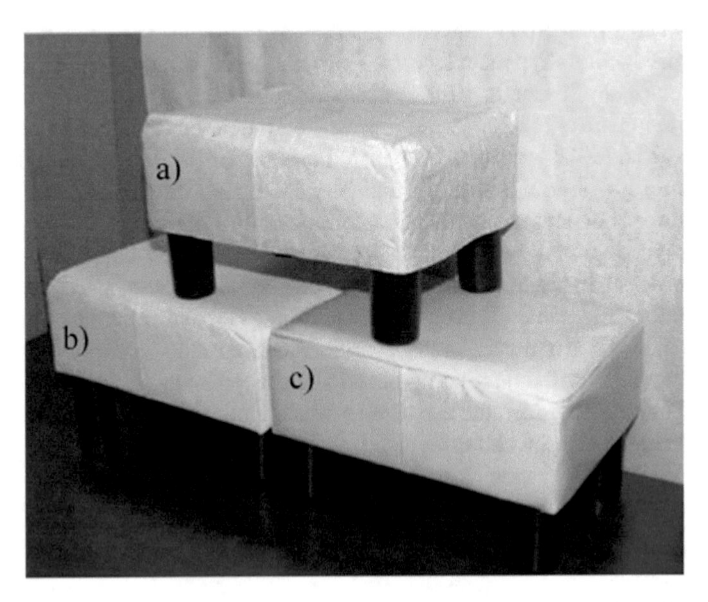

Fig. 4.12 Ottomans developed from **a** bacterial cellulose, **b** animal leather and **c** imitation leather [32]

4.4 Applications in Healthcare and Medical Industries

Bacterial cellulose is bestowed with wonderful properties such as a significantly higher amount of crystallinity in the range of 80–90% [33], water absorption capacity [34], and its degree of polymerisation is considerably high, i.e. up to 8000 [35]. Bacterial cellulose is biocompatible which is in addition to the above-mentioned properties enable it to be suitable to find its enormous applications in different sectors including healthcare and medical industries, in particular, those associated with biomedical and biotechnology applications [36]. Bacterial cellulose is widely applied in the medical and healthcare sector for artificial portable vessels, artificial leather, scaffolds for tissue engineering, wound healing materials and also for the drug delivery. In the biotechnology sector, bacterial cellulose is used as adsorbents, biosensors and biofilters for the immobilization of enzymes and cells and also as nutritional supplements. As bacterial cellulose is proved to be non-toxic by different scientists and as it possesses many desired properties for the medical field, it has accelerated for the medicinal development across the globe [1].

Bacterial cellulose is widely used in biomedical applications such as the following areas:

- Wound dressings
- Cornea replacement
- Medical implants
- Tissue engineering
- Antimicrobial products
- Drug delivery
- Soft tissue implant and cartilage replacements
- Wound healing
- Cancer targeting
- Cardiovascular implants
- Contact lenses.

Few of the most researched applications are detailed in this section of the chapter.

4.4.1 Wound Dressing Application

Among various potential applications of bacterial cellulose, one of the very first proposed applications of bacterial cellulose is related to wound dressing [37]. In 1990, Fontana et al. were the pioneers in describing the use of bacterial cellulose to replace burned skin [38]. Bacterial cellulose has many unique properties which makes it as an ideal scaffold choice for protecting injured tissues through wound dressing (especially for the applications as tissue regeneration, burn wounds and also as temporary skin substitutes). Demanding characteristics for these applications are: biocompatible, non-toxic and non-carcinogenic, moisture retention capability, capacity to absorb exudates from the injured tissue, and speed up granulation, and all these are possessed by bacterial cellulose. Hence, it is a suitable choice for wound dressing [39, 40]. Water holding and release capacity of bacterial cellulose due to its complex molecular structure with water molecules bonded through hydrogen bonds helps in increasing the healing capacity. Further, the capacity to shield the wounds from different infections, suppresses the pain and also helps to lower down the healthcare expenses. Additionally, the capability of water absorption and holding capacity will aid to charge liquid drugs and bioactive compounds on the wound dressing material. The desired ability to retain the required levels of humidity is of great help to elude the dehydration of the wound dressing and hence inhibits it from being attached to the wound, which is helpful in terms of shielding the tissue from exposure and reducing the pain throughout the entire dressing exchange process [41–43].

For temporary covering in the treatment of wounds, cellulose dressings are recommended including various treatments such as biopsy sites, tearing of the skin, venous stasis, ischaemic and diabetic wounds and pressure sores [44]. BioFill®, Bioprocess®, XCell® and Gengiflex® are some of the commercial bacterial cellulose-based wound dressings for periodontal diseases reconstruction [45]. The biomembrane BioFill® was one of the first commercial products for an ideal wound dressing which meets the desirable characteristics and requirements: low cost with durability, haemostatic, better handling, good elasticity, water vapour permeability, transparency, better adherence to the wound establish a physical barrier for bacteria, and very importantly this can be applied with a very limited number of exchanges. In addition to the above, the ability to accelerate the healing process and better relief of pain has been well proved in hundreds of cases [45–47]. It is significantly important to understand that the application of bacterial cellulose as a wound dressing has been proved clearly that it has reduced the required healing time, which is the ultimate need for a wound dressing which was noted and reported by many researchers [46, 48, 49].

Studies showed that bacterial cellulose is superior to conventional wound dressings in terms of exudate retention, reducing wound pain, accelerating re-epithelialization and healing times, reducing wound infection rates, ease of wound inspection (which is due to the semi-transparency of BC), and in reducing scarring [50]. The gelatinous membrane of bacterial cellulose is usable as a temporary skin substitute for the treatment of wounds like pressure sores, skin tears, venous stasis, ischaemic and diabetic wounds, second-degree burns, skin graft donor sites, traumatic abrasions and lacerations, ulcers, grafts because of its high mechanical strength in the wet state, substantial permeability for liquids and gases and low irritation of skin [51, 52]. A moist environment facilitates the penetration of active substances into the wound and enables easy and painless dressing change without damage to the newly formed skin [53]. BC can provide a moist environment to the wound site and it can easily conform to the contour of the wound thus providing a barrier between the wound and the surrounding environment [54]. The bacterial cellulose modified with chitosan had high elongation at the break that shows the good elasticity which makes it fits the wound site well and facilitates protection against external infection [53]. Bacterial cellulose can maintain a proper water balance by either absorbing or releasing fluid according to the behaviour of the wound [50]. The other important factor that affect the wound healing process is the infection caused by microorganisms present in the environment. The application of wound dressing materials with good antibacterial and barrier properties against microorganisms may be a good solution to this problem [53].

Nanocomposites are one of the recent developments which gained noteworthy attention and owing to the excellent characteristics possessed by them, they are used in many applications. It is well known that composites consist of two types of individual materials—the matrix and the reinforcement material. The matrix acts as a scaffold and also supports the reinforcement material and reinforcements are responsible to deliver the physical, chemical and biological properties to the matrix. Bacterial cellulose can be used as both the matrix and scaffold for the composite design. Owing to the desired properties such as high specific surface area, highly porous structure and mechanical strength possessed by bacterial cellulose, it is found to be the best substrate to house a range of nanomaterials [55]. Manipulation of bacterial cellulose by incorporating improvements and several modifications to form nanocomposites to enhance its capabilities with improved properties and better functionalities is the next level application. Three different approaches have been used to incorporate nanomaterials into/onto the BC matrix, which are:

(i) directly adding of nanomaterials into a BC matrix
(ii) synthesis of nanomaterial in the structure of BC-based materials and
(iii) direct coating of nanomaterials as a nanolayer on the BC surface. So far, numerous nanomaterials, such as metal and metal oxide nanoparticles (Ag, Au, Ni, Pd, CuO and TiO_2), mineral nanomaterials (SiO_2, $CaCO_3$, montmorillonite) and carbonaceous nanomaterials (graphene, carbon nanotube) have been stored into nanocellulose matrices for the preparation of BC nanocomposites [55].

Bacterial cellulose pellicles can be impregnated with silver particles which can impart its antimicrobial properties [54]. Bacterial cellulose-Ag nanocomposites were found to be effective against many bacterial and fungal species, thereby reducing the chances of wound infection when utilised as dressing materials. Strong antimicrobial activity is exhibited against gram-positive and gram-negative bacteria by the bacterial cellulose nanocomposites impregnated with freeze-dried silver nanoparticles. Propolis, being a natural substance with good antifungal, antiviral, antioxidant, anti-inflammatory and antibacterial properties is incorporated in the surface and interstices of bacterial cellulose to form propolis-bacterial cellulose composites which favour tissue repair in less time and more effectively in contaminated wounds [52]. And also, a wide variety of bacterial cellulose composites are being used in wound dressing applications. Bacterial cellulose with chitosan, poly(3-hydroxybutyrate-co_4-hydroxybutyrate), kaolin, gelatin and collagen composites are proved to have better wound healing properties than native cellulose [54]. Bacterial cellulose incorporated with alginates is being used to prepare a unique type of sponges that are having a good potential to be used to cover the surgical wounds of the oral cavity [54]. The healing mechanism of the wound using cellulose-based wound dressing materials involves the capture of ions by means of cellulose hydrogen bonds or the nanobacterial cellulose 3D network mimics the skin surface and provides optimal healing conditions. Yet, the mechanism is not fully understood. It is observed that the meantime for 75% epithelization was reduced from 315 days to 81 days while using bacterial cellulose-based wound dressing on the chronic wounds [52].

4.4.2 Cardiovascular Implants

Bacterial cellulose has been investigated as an artificial blood vessel because of its processing properties [52]. Mouldability of bacterial cellulose gives it the advantage of synthesising in the form of regular tubes with a different inner diameter, wall thickness and length. The technique of moulding it during fermentation is patented [3]. Bacterial cellulose tubes have good surgical handling and can be sterilized in standard ways. The tubes presented excellent mechanical properties and hold promise for use as a microvessel or soft tissue material in medical and pharmaceutical applications. The load capacities of the traverse section of BC are comparable with the rat blood vessels. In a follow-up in vivo study with rats, pigs and sheep, these tubes were successfully used to replace carotid arteries [3, 11]. Nanocomposites made of PVA and bacterial cellulose could exhibit a broad range of mechanical properties, especially anisotropy which are essential for vascular grafts [3]. This biocompatible composite possessed good mechanical stress-strain characteristics which are comparable to native heart leaflets. Currently, the available standard valve replacements are either porcine heart valves or mechanical heart valves. Porcine heart valves have the disadvantage of a limited lifespan, whereas mechanical heart valves last longer but they require lifelong blood-thinning medication and cause increased shear forces on red blood cells, cell rupture and associated anaemia. These limitations of current treatments made bacterial cellulose composite material as an alternative [4] and also the mechanical properties of bacterial cellulose are comparable to the porcine carotid artery and better than expanded polytetrafluoroethylene [3].

4.4.3 Cartilage/Meniscus Implants

The regeneration capacity of cartilage tissue is limited and repairing cartilages is a challenging one [52]. Bacterial cellulose/poly (dimethyl acrylamide) double network gel has superior properties that could be helpful in using cellulose as a cartilage substitute. Bacterial cellulose composites surpassed alginate and plastic in supporting chondrocyte migration and proliferation and it reveals its potential utility as a cartilage scaffold. Unmodified bacterial cellulose is similar to porcine knee menisci and superior to collagen-based materials in Young's modulus but still inferior to natural menisci at higher loads. However, the characteristics can be improved by directing bacterial cellulose fibril orientation and through composites by impregnating into proteoglycans and chondrocytes [50]. Bacterial cellulose modified by the addition of *n*-acetyl glucosamine residues in the HS medium had a higher level of mesenchymal stem cell proliferation than that of native bacterial cellulose. For human auricular cartilage replacement, bacterial cellulose with 15% cellulose content has been proposed as it matches the mechanical strength [54].

4.4.4 Drug Delivery Systems

Bacterial cellulose is one such biopolymer that fulfils the criteria for consideration as a drug delivery material. In recent years, for pharmaceutical applications, nanocellulose material-based drug delivery systems are being proposed. It is investigated as a potential drug delivery system for proteins. Moreover, freeze-dried material showed a lower uptake of protein than the pristine. Using spray coating techniques, tablets are film-coated with bacterial cellulose and their drug release characteristics are studied. Bacterial cellulose is a material with the ability to form soft, flexible and foldable films that provided appropriate drug delivery [52].

4.4.5 Scaffolds

Scaffolds are a novel system that is used to support organs and organ systems that may have been damaged after injury or disease. Scaffolds closely mimic the complex and hierarchical structures inherent to the native tissue being designed to provide the microenvironment that cells need to proliferate, migrate and differentiate. For promoting an ideal environment to host cells, hydrogel-like properties are essential. Bacterial nanocellulose network has a very high affinity for water which results in such hydrogel-like properties. Various studies have confirmed that in the presence of bacterial cellulose scaffold, cells such as human embryonic kidney cells, bone-forming osteoblasts, and fibroblasts, human smooth muscle cells can be grown. Graphene oxide-bacterial cellulose nanocomposite hydrogels are seemed to have better tensile strength and Young's modulus than pure bacterial cellulose. These hydrogels are emerging as a new material for tissue engineering scaffolds. A bacterial cellulose-alginate scaffold composite with biocompatibility and open macro-porous structure could be potentially used as a scaffold for tissue engineering. Hep-bacterial nanocellulose nanofibrous scaffolds which are prepared by hybridization of heparin with bacterial cellulose network showed good anticoagulant properties for the potential use in vascular tissue engineering. Bacterial cellulose-hydroxyapatite scaffolds obtained showed an excellent results in terms of regeneration of bone and connective tissues [52].

4.4.6 Bone and Connective Tissue Repair

Bone is a composite material comprising an organic phase (collagen and non-collagenous proteins) and an inorganic mineral phase (calcium hydroxyapatite). Bacterial cellulose is a promising matrix for obtaining different types of calcium carbonate crystals along with improved biocompatibility. Potential bone regeneration is achieved by a membrane made of bacterial cellulose and hydroxyapatite. It delivered prone growth of osteoblast cells, a high level of alkaline phosphatase activity, and greater bone nodule formation. This bacterial cellulose-hydroxyapatite membranes were suitable for bone regeneration and accelerated new bone formation. Induction of negative charge on cellulose by the adsorption of polyvinylpyrrolidone initiated the nucleation of hydroxyapatite through dynamic simulated body fluid treatment. A novel bone repair biomaterial is developed by the introduction of goat bone apatite in bacterial cellulose as it can stimulate bone cell proliferation and promote cell differentiation [52].

4.4.7 Dental and Oral Implants

Alginate-based bacterial cellulose composites showed a good result while using as a temporary dressing for the surgical wounds of the oral mucosa. While comparing with the conventional paper point materials used for the dental root canal treatment, bacterial cellulose exhibited greater compatibility and biological characteristics. The properties like better adsorption rate and better tensile behaviour in the wet state than that of paper point materials made bacterial cellulose as an innovative material for dental root canal treatment [52].

4.4.8 Neural Implants

The reconstruction of nervous tissue is a challenging problem and the difficulty of self-regeneration depends on the extent of harm. Nerve growth factor neurotrophin is proliferated by the bacterial cellulose to which mesenchymal stem cells adhere. It could create a microenvironment that promotes neuronal regeneration. Bacterial cellulose neurotubes can effectively prevent the formation of neuromas, while allowing the accumulation of neurotrophic factors inside, and facilitating the process of nerve regeneration. The biocompatibility of bacterial cellulose with Schwann cells made it useful in the preparation of nerve conduits for repairing peripheral nerve injuries and they showed no adverse haematological and histological effects upon implantation in rats [52].

4.4.9 Artificial Cornea/Contact Lens

A special mechanism is patented to conform bacterial cellulose to specific shapes and angles to apply it in the field of cornea regeneration. The contact lens-shaped membrane is made by cutting the wet cellulose membrane in a round shape and compressed at 150 °C heat with the stick having a semi-spherical end. The therapeutic potential of bacterial cellulose membrane in the shape of a contact lens is improved by the impregnation of ciprofloxacin with and without 2-hydroxypropyl-cyclodextrin into cellulose membrane. The membranes did not exhibit cytotoxicity, genotoxicity or mutagenicity effects and this made it a promising biomaterial with the application of contact lens used for regeneration and protection against bacteria [52].

4.5 Other Engineering Applications

Other than the medical and clothing application, there are few other explored applications of bacterial cellulose is listed in this section to report the potentiality of the bacterial cellulose. A few of the most common applications are discussed in the following sections.

4.5.1 Food Applications

From the earlier century, bacterial cellulose is used for food application as a dessert, Nata de coco. Due to the low calories, it is mainly used as a food supplement and also used a meat substitute. In 1992, the FDA classified the bacterial cellulose as a safe component for food [56]. Generally, bacterial cellulose paste is used in the food product as a viscosity controlling agent. In chocolate drinks, the addition of bacterial cellulose reduces the viscosity compared to the xanthan gum. A similar application also reported by the researcher for ice creams. The addition of bacterial cellulose in ice creams reduces the flow after the melting of the ice cream. The use of bacterial cellulose generally maintains the humidity of the food product for a longer duration. This helps the food to be stored for a longer period of time. For instance, the researchers mentioned that the ice creams with cellulose retain its contour for 60 min after the removal of ice creams from the freezer. The absence of the cellulose will reduce this time leads to faster melting [57].

The application of bacterial cellulose widely requires in food industry whenever there are requirements in lack of flavour interactions, foam stabilization and stability over a wide pH range, temperature and freeze-thaw conditions [58]. Many researchers analysed the use of bacterial cellulose as a cell immobilizing agent. Due to its higher surface area, mechanical strength and biocompatibility, bacterial cellulose noted as a suitable agent for cell immobilization. A research report mentioned that the use of

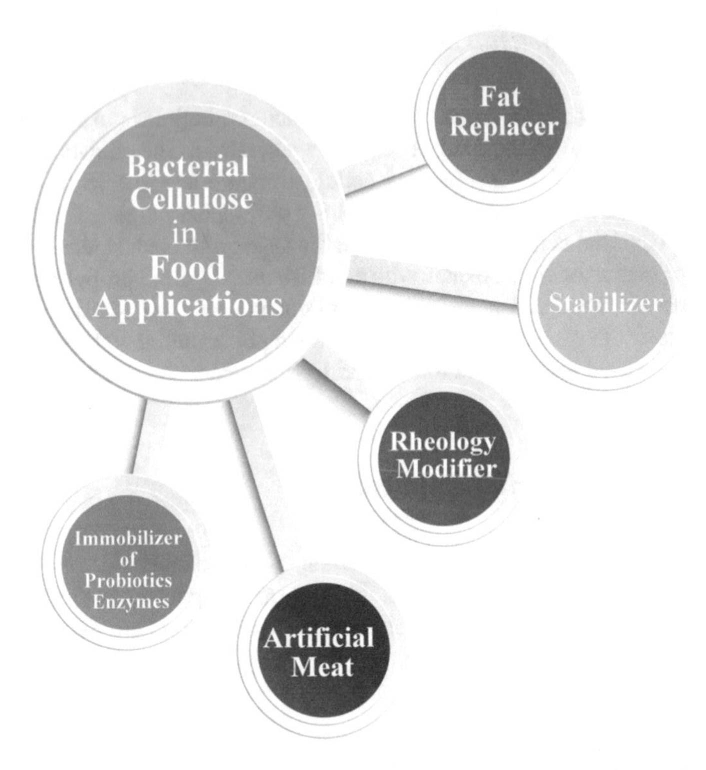

Fig. 4.13 Bacterial cellulose applications in the food industry [Authors own illustration]

bacterial cellulose in the wine fermentation process increased the metabolic activity of the yeast [59]. The material also used as a fat replacer in several foods as the addition of fat causes several health issues to humans. The addition of 20% bacterial cellulose in the meatballs by completely removing the fat content resulted in a change in the softening, cooking loss and product acceptance. A similar effect also noted for the 10% addition as reported by the researchers [60]. Several other applications of bacterial cellulose in the food industries are listed in Fig. 4.13.

4.5.2 Filtration Applications

Eco-friendly nature and a highly porous structure with superior mechanical properties of the bacterial cellulose is one of the main reasons for its applications in filtration area. No much work found under the filtration area, however, few interesting studies are listed here. A research work analysed the use of bacterial cellulose and cellulose nanofibre membrane for filtration of an oil-in-water emulsion. The researchers reported the harvesting time did not have any significant impact on the porosity of the bacterial cellulose and so the filtration properties. The filtration efficiency of wet bacterial cellulose is noted higher than the cellulose nanofibrils. This is mainly attributed to the never dried structure of the bacterial cellulose [61]. Various components have been included in the bacterial cellulose non-woven to enhance the filtration properties. The addition of graphene oxide showed increased mechanical strength and selective ion permeation to the angstrom scale due to the graphene addition. The researchers reported that the composite showed higher potential for the water and pharma industry purification process [62]. On the analysis of pure water filtration efficiency, it is noted that the increment in the cultivation time reduced the efficiency from 20.5 to 4.5 L/m^2 in contrast to the previous researcher [61]. The increment in the cultivation time reduced the porous nature of the structure. On the analysis of the textile wastewater filtration efficacy using bacterial cellulose, the results reported that pH of the effluent played a major role in colour and chemical oxygen demand (COD) removal. The reduction of the effluent pH from 8 to 5 increased the filtration efficiency of colour from 51.6 to 85.9% and 49.2 to 73.8% in the case of COD removal. Based on the results, the researchers suggested the bacterial cellulose as a potential filtration media for textile effluent treatment [63].

Takai evaluated the filtration properties of the bacterial cellulose with several other polymers like polyethylene glycol (PEG), CMC, carboxymethyl chitin and other cellulose-based polymers. The researcher developed a composite membrane and evaluated the filtration performance of the composite and reported that the bacterial cellulose had the potential of being an ultra-filtration membrane [64]. Recently, few researchers tried to use raw kombucha-based bacterial cellulose for filtration applications and they call it a live filtration membrane. In their study, the researchers analysed the pure water permeability test after one-hour compaction. The results reported a permeability value (average) of 135 ± 25 L m^{-2} h^{-1} bar^{-1} noted for different applied pressures ranging from 0.7 to 3.1 bar. But in the same condition, a traditional polymeric ultra-filtration system can produce approximately 1000 L m^{-2} h^{-1} bar^{-1}. However, the potentiality of the bacterial cellulose membrane can be explored if different composite structures and filtration pressures were evaluated. The researcher also proposed a day-to-day application of the developed living membrane for the production of purified drinking water. As the membrane can remove nano-size organisms and dust, the filtration aids to get pure drinking water. The experimental setup proposed by the researcher was provided in Fig. 4.14. The researchers reported that filtration of 300 ml of water obtained after 8 h of filtration without any additional pressure. However, the researcher mentioned that the flow rate can be increased by applying gravity pressure using an elevated water tank setup [65].

4.5.3 Electrical and Sensor Applications

Superior mechanical and structural properties of bacterial cellulose have unlimited potential in the electrical and biomedical sensor applications. Generally, the bacterial cellulose is functionalised with a different conductive polymeric material and used in the electrical applications. Bacterial cellulose material was successfully used in the

Fig. 4.14 Point of use application with potential operational setup for gravity-based bacterial cellulose filtration [65] [Reprinted with Permission]

gas sensing application as a sensor. To utilise the flexible, nanoporous and nanofibril effectively in the electrical field, the researcher used amino-functionalized graphene to functionalise the bacterial cellulose. Through simple single-step esterification, the bacterial cellulose- amino-functionalized graphene composites developed. Later on, the crosslinking of conductive polymer polyaniline, bacterial cellulose-based composites potential on gas sensing applications are analysed [66]. To address the biodegradability issue of electronic components in the afterlife state, the researcher aimed to develop paper-based mass sensors for health monitoring and ambient assisted living. The researcher developed a bacterial cellulose-based composite using the conducting polymer. The developed composite is treated with ionic liquid and its application potential as a mass sensor is measured. The results were positive and produced a sensitivity of −28.6 Hz/g. The sensor was able to develop a frequency resolution equal to 0.1 Hz. With the results, the researchers can be able to measure a mass of 3.5 mg. The results were in good agreement between the theoretical estimation of the study. The use of bacterial cellulose provides a sustainable material for the electrical application [67]. Di Pasquale et al. characterised the bacterial cellulose treated with ionic liquid and coated with conducting polymers for accelerometer applications. The developed system works in the cantilever configuration. The findings of the study reported both the chemical and mechanoelectrical characterisation of the developed bacterial cellulose-based devices. The researcher proposed the organic and bionature of the bacterial cellulose as a major advantage of the developed composite over the competing technologies in the market [68].

Next to the sensors, bacterial cellulose materials are also used as an electrode in the fuel cells. Recent research reported the potentiality of the bacterial cellulose to catalyse the precipitation of palladium within its structure. It was reported that the three-dimensional network of fibres in the structure, and the higher surface area is the important reason for the higher catalytic potential of the bacterial cellulose. They also reported the ability of bacterial cellulose to promote the precipitation of metals like palladium, gold and silver from the aqueous solution. After the precipitation, the metalized bacterial cellulose materials can be used as anodes or cathodes in biosensors and fuel cells [69]. Other researchers used different materials along with bacterial cellulose to develop sensors. They used metal nanoparticles, metal oxides, carbon-based materials (carbon nanotubes, graphene, carbon dots), conductive polymers and enzymes in the second step functionalisation. The researcher proposed two different approaches in their work namely developing sensors without spoiling the structure of the bacterial cellulose and by spoiling it. In the first case, the cellulose 3D structure is preserved by either in-situ or ex-situ application of the functional chemical. In the latter case, by converting the bacterial cellulose as nanocrystals or fibres, the sensors have been developed. In their research, they proposed several applications of the bacterial cellulose sensors based on the type of signal they work with as resonant, optical or electrochemical [70]. Other research works proposed a bacterial cellulose-carbon paste electrode. They developed a novel composite using bacterial cellulose and carbon. The nanostructure and lower cost are the major advantages proposed by the researcher. The higher electrochemical property of the bacterial cellulose-based carbonaceous materials provides a very huge potential for bacterial cellulose as paste electrodes [71]. Several other application potentials of bacterial cellulose in the electrical sensor fields as listed by the previous researcher are (i) humidity biosensor applications, (ii) biomedical biosensor applications and (iii) pollution biosensor applications [70].

4.5.4 Paper Industry

The higher tensile properties of the bacterial cellulose are found to be one of the major reasons for its usage in the paper industry. The earlier time Yamanaka and Watanabe used the bacterial cellulose in the papermaking process. Their study found that the addition of bacterial cellulose increased the strength of the paper. The reinforcement of bacterial cellulose and paper pulp increased the folding endurance of the paper. A 15% addition of bacterial cellulose paper composite showed a four-time higher folding strength to the paper and Young's modulus increased from 2 to 3.5 Gpa [72]. In a different study, the researcher compared the properties of the papers produced in bacterial cellulose composite with static and agitated culture. The researchers used bacterial cellulose pulp along with the recycled old paper pulps. The results reported that the addition of agitated bacterial cellulose increased the tensile and tear properties of the developed paper to 12.2 and 14.2%. The results were different from static cultured bacterial cellulose. The addition of static culture increased the tear strength but it does not influence the tensile properties. This is due to the nature of the static culture process, the bacterial cellulose grows on the top of the vessel where the paper pulp sediment at the bottom during the cultivation [73]. Other researchers used the phosphate along with glucose during the cultivation of the bacterial cellulose. The researcher added the phosphate bacterial cellulose and native bacterial cellulose in the papermaking process. The results reported that the addition of phosphate bacterial cellulose in the paper increased the kaolin retention, strength and fire resistance properties as compared to the normal bacterial cellulose. The research work demonstrated the development of functional paper using bacterial cellulose [74].

Other studies developed fluorescent paper by incorporating the bacterial cellulose and fluorescent particles inside the paper. In their research, they first developed a composite of bacterial cellulose by incorporating with Europium, one of the best photoluminescents. By mixing the complex with plant-based cellulose, they developed fluorescent paper with higher fluorescent nature and illumination property. They had reported that 5% bacterial cellulose complex provided better fluorescent nature and further addition did not show any changes. The paper possessed higher durability even after folding 200 times there was only a 0.7% reduction in fluorescent efficiency noted [75]. The use of dispersed bacterial cellulose modified by polypropylene with plant fibres increased the mechanical properties of the paper to a greater extent. The research reported for an addition of 3% modified bacterial cellulose, the developed paper showed the following changes in the physical properties. A 12.6% increase in the tensile index, a 10.1% increment in the paper's tear index, and a 7.82% increment in the bursting index over the conventional one. The addition of bacterial cellulose bridges the gap between the plant cellulose and so developed a tremendous increment in the physical properties of the paper [76].

4.5.5 Other Applications

Other than the above-mentioned applications, bacterial cellulose and its complex forms have different application areas as listed

1. Food and other packaging applications
2. Cosmetics
3. Speaker diaphragms
4. Semiconductors
5. Adsorbents
6. Composites
7. Pharmaceutical applications
8. Aerogels
9. Energy production (bioethanol).

4.6 Summary

The various applications of bacterial cellulose are detailed in this chapter. The main emphasis is given on the apparel and fashion end use of the bacterial cellulose as a clothing, leather alternative, as fashion accessories, etc. Several possibilities of the applications detailed. Medical and healthcare applications of the bacterial cellulose is a well-explored area and so the other part enlightens application areas in the region. The other engineering applications like paper industry, sensors and electrical application, filtration and food industrial application possibilities are also detailed in the chapter.

References

1. Niyazbekova ZT, Gulden N, Askar K (2018) An overview of bacterial cellulose applications. In: Biotechnology theory and practice. https://doi.org/10.11134/btp.2.2018.3
2. Brown JRM (1992) Emerging technologies and future prospects for industrialization of microbially derived cellulose, in harnessing biotechnology for the 21st century. In: Ladisch MR, Bose A (eds) Proceedings of the 9th international biotechnology symposium and exposition. American Chemical Society, Virginia, Washington, pp 76–79
3. Mohammedi Zohra (2017) Structure, properties and medical advances for biocellulose applications: a review. Am J Polymer Sci Technol 3(5):89–96. https://doi.org/10.11648/j.ajpst.20170305.12
4. Bielecki S, Krystynowicz A, Turkiewicz M, Kalinowska H (2002) Bacterial cellulose. In: Vandamme J, Baets SD, Steinbüchel A (eds) Biopolymers: polysaccharides I: polysaccharides from prokaryotes, vol 5. Wiley-VCH Verlag, Weinheim, pp 37–90
5. The Mag, Making Clothes from Microbes (2014) https://www.mentalfloss.com/article/55269/making-clothes-microbes. Accessed 1 Aug 2020
6. Venkataramanan M (2014) Biocouture from the lab to the high street. https://www.wired.co.uk/article/biocouture-from-the-lab-to-the-high-street. Accessed 1 Aug 2020
7. BioCouture by Suzanne Lee (2011) https://www.iconeye.com/design/news/biocouture-by-suzanne-lee. Accessed 1 Aug 2020
8. Ng MCF, Wang W (2015) A study of the receptivity to bacterial cellulosic pellicle for fashion. Res J Text Appar 19:65–69
9. Ng FMC, Wang PW (2016) Natural self-grown fashion from bacterial cellulose: a paradigm shift design approach in fashion creation. Des J 19:837–855
10. da Silva FM, Gouveia IC (2015) The role of technology towards a new bacterial-cellulose-based material for fashion design. J Ind Intell Inf 3(2)
11. Jennifer H, Natalie T, Logan F (2017) Durability properties of bacterial cellulose for textile applications. In: International textile and apparel association (ITAA) annual conference proceedings. 166. https://lib.dr.iastate.edu/itaa_proceedings/2017/posters/166
12. Jennifer H, Logan F, Natalie T (2017) Comfort and aesthetic properties of bacterial cellulose for textile applications. In: International textile and apparel association (ITAA) annual conference proceedings. 6. https://lib.dr.iastate.edu/itaa_proceedings/2017/presentations/6
13. Han Juyeon, Shim E, Kim HR (2019) Effects of cultivation, washing, and bleaching conditions on bacterial, cellulose fabric production. Text Res J 89(6):1094–1104
14. Shim E, Kim HR (2019) Coloration of bacterial cellulose using in situ and ex situ methods. Text Res J 89(7):1297–1310
15. Devon W, Hang L, Salusso CJ (2015) Production and characterization of bacterial cellulose fabrics. In: International textile and apparel association (ITAA) annual conference proceedings. 143. https://lib.dr.iastate.edu/itaa_proceedings/2015/posters/143

16. Tyurin I, Getmantseva V, Andreeva E, Kashcheev O (2019) The study of the molding capabilities of bacterial cellulose. In: AUTEX2019—19th world textile conference on textiles at the crossroads, 11–15 June 2019, Ghent, Belgium. https://ojs.ugent.be/autex/article/view/11745
17. Jennifer H (2017) Homegrown: investigating design potential of bacterial cellulose. In: International textile and apparel association (ITAA) annual conference proceedings. 15. https://lib.dr.iastate.edu/itaa_proceedings/2017/design/15
18. Chan CK, Shin J, Kinor Jiang SX (2018) Development of tailor-shaped bacterial cellulose textile cultivation techniques for zero-waste design. Clothing Text Res J 36(1):33–44
19. Domskiene J, Sederaviciute F, Simonaityte J (2019) Kombucha bacterial cellulose for sustainable fashion. Int J Clothing Sci Technol 31(5):644–652. https://doi.org/10.1108/IJCST-02-2019-0010
20. Kaminski Kamil, Jarosz Magdalena, Grudzien Joanna, Pawlik Justyna, Zastawnik Filip, Pandyra Piotr, Kołodziejczyk AM (2020) Hydrogel bacterial cellulose: a path to improved materials for new eco-friendly textiles. Cellulose 27:5353–5365. https://doi.org/10.1007/s10570-020-03128-3
21. Mihaleva G, Bio matter in creative practises for fashion and design, AI & SOCIETY. https://doi.org/10.1007/s00146-020-00957-5
22. Audrey N (2017) Grown microbial 3D fiber art, ava: fusion of traditional art with technology. ISWC '17, 11–15 Sept 2017, Maui, HI, USA
23. Yim SM, Song JE, Kim HR (2017) Production and characterization of bacterial cellulose fabrics by nitrogen sources of tea and carbon sources of sugar. Process Biochem 59:26–36
24. Armine G (2018) Evaluation of consumer perceptions and acceptance of sustainable fashion products made of bacterial cellulose. Graduate Theses and Dissertations. 16583. https://lib.dr.iastate.edu/etd/16583
25. Armine G (2017) Made from scratch. A sustainable handbag made of bacterial cellulose grown in fermenting tea. In: International textile and apparel association (ITAA) annual conference proceedings. 65. https://lib.dr.iastate.edu/itaa_proceedings/2017/design/65
26. Fernandes M, Gama M, Dourado F, Souto AP (2019) Development of novel bacterial cellulose composites for the textile and shoe industry. Microb Biotechnol 12:650–661
27. Changhyun N, Lee YA (2016) RETHINK II: Kombucha Shoes for Scarlett and Rhett. In: International textile and apparel association (ITAA) annual conference proceedings. 68. https://lib.dr.iastate.edu/itaa_proceedings/2016/design/68
28. The Making of Nata de Coco Shoes. Available online: http://avrotor.blogspot.com/2017/12/the-making-ofnata-de-coco-shoes.html. Accessed 28 Feb 2020
29. Hu W, Chen S, Yang J, Li Z, Wang H (2014) Functionalized bacterial cellulose derivatives and nanocomposites. Carbohydr Polym 101:1043–1060
30. Stephen R, Coconut water in your shoe. https://www.thehindu.com/life-and-style/homes-and-gardens/coconut-water-in-your-shoe/article25742867.ece. Accessed 27 June 2020
31. Payne A (2016) Will we soon be growing our own vegan leather at home? https://www.dailybulletin.com.au/the-conversation/24701-will-we-soon-be-growing-our-own-vegan-leather-at-home. Accessed 27 Feb 2020
32. Solatorio N, Chong Liao C (2019) Synthesis of Cellulose by *Acetobacter Xylinum*: a comparison vegan leather to animal and imitation leather. Honors thesis, University of Wyoming, Spring
33. Keshk SM (2014) Bacterial cellulose production and its industrial applications. J Bioprocess Biotech 4:150. https://doi.org/10.4172/2155-9821.1000150
34. Saibuatong O, Phisalaphong M (2010) Novo aloe vera–bacterial cellulose composite film from biosynthesis. Carbohydr Polym 79:455–460. https://doi.org/10.1016/j.carbpol.2009.08.039
35. Dahman Y (2009) Nanostructured biomaterials and biocomposites from bacterial cellulose nanofibers. J Nanosci Nanotechnol 9:5105–5122. https://doi.org/10.1166/jnn.2009.1466
36. Moniri M, Boroumand Moghaddam A, Azizi S et al (2017) Production and status of bacterial cellulose in biomedical engineering. Nanomaterials (Basel) 7(9):257. Published 4 Sept 2017. https://doi.org/10.3390/nano7090257
37. Helida B, Robson S, Hernane B, Agnieszka T, Junkal G, Wilton L, Osmir O Jr, Sidney R (2016) A multipurpose natural and renewable polymer in medical applications: bacterial cellulose. Carbohydr Polym 153:406–420. https://doi.org/10.1016/j.carbpol.2016.07.059

38. Fontana JD, De Souza AM, Fontana CK, Torriani IL, Moreschi JC, Gallotti BJ et al (1990) Acetobacter cellulose pellicle as a temporary skin substitute. Appl Biochem Biotechnol 24–25:253–264
39. Khalid A, Ullah H, Ul-Islam M, Khan R, Khan S, Ahmad F et al (2017) Bacterial cellulose-TiO_2 nanocomposites promote healing and tissue regeneration in burn mice model. RSC Adv 7:47662–47668
40. Li Y, Jiang H, Zheng WF, Gong NY, Chen LL, Jiang XY, Yang G (2015) Bacterial cellulose-hyaluronan nanocomposite biomaterials as wound dressings for severe skin injury repair. J Mater Chem B 3:3498–3507
41. Agarwal A, McAnulty JF, Schurr MJ, Murphy CJ, Abbott NL (2011) Polymeric materials for chronic wound and burn dressings. In: Farrar D (ed) Advanced wound repair therapies. Elsevier Inc., Amsterdam, The Netherlands, pp 186–208
42. Shah N, Ul-Islam M, Khattak WA, Park JK (2013) Overview of bacterial cellulose composites: a multipurpose advanced material. CarbohydrPolym 98:1585–1598
43. Ovington LG (2007) Advances in wound dressings. Clin Dermatol 25:33–38
44. Kowalska-Ludwicka K, Cala J, Grobelski B, Sygut D, Jesionek-Kupnicka D, Kolodziejczyk M et al (2013) Modified bacterial cellulose tubes for regeneration of damaged peripheral nerves. Arch Med Sci 9:527–534
45. Farah LFX (1990) Process for the preparation of cellulose film, cellulose film produced thereby artificial skin graft and its use. Bio Fill Produtos biotecnológicos (Brasil). US 4912049, 30 Sept 1990, 10 June 1986
46. Czaja W, Young DJ, Kawecki M, Brown RM Jr (2007) The future prospects of microbial cellulose in biomedical applications. Biomacromol 8:1–12
47. Wouk AFPF, Diniz JM, Círio SM, Santos H, Baltazar EL, Acco A (1998) MembranabiológicaBiofill-estudocomparativo com outros agentespromotores da cicatrizac¸ão da peleemsuínos: Aspectosclínicos, histopatológicos e morfométricos. Arch Vet Sci 3:31–37
48. Czaja W, Krystynowicz A, Bielecki S, Brown RM Jr (2006) Microbialcellulose-the natural power to heal wounds. Biomaterials 27:145–151
49. Portal O, Clark WA, Levinson DJ (2009) Microbial cellulose wound dressing in the treatment of nonhealing lower extremity ulcers. Wounds 21:1–3
50. Petersen N, Gatenholm P (2011) Bacterial cellulose-based materials and medical devices: current state and perspectives. Appl Microbiol Biotechnol 91:1277–1286. https://doi.org/10.1007/s00253-011-3432-y
51. Sherif MAS Keshk (2014) Bacterial cellulose production and its industrial applications. J Bioprocess Biotech 4(2). https://doi.org/10.4172/2155-9821.1000150
52. de Oliveira Barud HG, da Silva RR, da Silva Barud H, Tercjak A, Gutierrez J, Lustri WR, de Oliveira Jr OB, Ribeiro SJL (2016) A multipurpose natural and renewable polymer in medical applications: bacterial cellulose. Carbohydr Polym. http://dx.doi.org/10.1016/j.carbpol.2016.07.059
53. Ciechańska D (2004) Multifunctional bacterial cellulose/ chitosan composite materials for medical applications. Fibres Text East Eur 12(4) (48):69–72
54. Rajwade JM, Paknikar KM, Kumbhar JV (2015) Applications of bacterial cellulose and its composites in biomedicine. Appl Microbiol Biotechnol. https://doi.org/10.1007/s00253-015-6426-3
55. Moniri M, Boroumand Moghaddam A, Azizi S, Abdul Rahim R, Bin Ariff A, Zuhainis Saad W, Navaderi M, Mohamad R (2017) Production and status of bacterial cellulose in biomedical engineering. Nanomaterials 7(9):257. https://doi.org/10.3390/nano7090257
56. Ullah H, Santos HA, Khan T (2016) Applications of bacterial cellulose in food, cosmetics and drug delivery. Cellulose 23(4):2291–2314
57. Shi Z, Zhang Y, Phillips GO, Yang G (2014) Utilization of bacterial cellulose in food. Food Hydrocoll 35:539–545
58. Okiyama A, Shirae H, Kano H, Yamanaka S (1992b) Bacterial cellulose I. Two-stage fermentation process for cellulose production by *Acetobacter aceti*. Food Hydrocol 6(5):471–477

59. Ton NMN, Le VVM (2011) Application of immobilized yeast in bacterial cellulose to the repeated batch fermentation in wine-making. Int Food Res J 18(3):983–987
60. Lin KW, Lin HY (2004) Quality characteristics of Chinese-style meatball containing bacterial cellulose (Nata). J Food Sci 69:Q107–Q111. https://doi.org/10.1111/j.1365-2621.2004.tb13378.x
61. Hassan Enas, Hassan Mohammad, Abouzeid Ragab, Berglund Linn, Oksman K (2017) Use of bacterial cellulose and crosslinked cellulose nanofibers membranes for removal of oil from oil-in-water emulsions. Polymers 9:388. https://doi.org/10.3390/polym9090388
62. Fang Q, Zhou X, Deng W et al (2016) Freestanding bacterial cellulose-graphene oxide composite membranes with high mechanical strength for selective ion permeation. Sci Rep 6(1):33185
63. Isik Zelal, Unyayar Ali, Dizge Nadir (2018) Filtration and antibacterial properties of bacterial cellulose membranes for textile wastewater treatment. Avicenna J Environ Health Eng 5(2):106–114
64. Takai M (1994) Bacterial cellulose composites. In: Gilbert RD (ed) Cellulose polymer blends composites. Hanser, Munich, pp 233–240
65. Eggensperger CG, Giagnorio Mattia, Holland Marcus C, Dobosz Kerianne M, Schiffman Jessica D, Tiraferri Alberto, Zodrow Katherine R (2020) Sustainable living filtration membranes. Environ Sci Technol Lett 7:213–218
66. Abdali Hana (2019) Synthesis of graphene/bacterial cellulose/polyaniline nanocomposite for gas detection. Am J Adv Drug Deliv 7:53
67. PasqualeGD, Graziani S, Pollicino A, Trigona C (2019) A bacterial cellulose based mass sensor. In: 2019 IEEE international symposium on measurements & networking (M&N), Catania, Italy, 2019, pp 1–4. https://doi.org/10.1109/IWMN.2019.8805008
68. Di Pasquale G, Graziani S, Pollicino A, Trigona C (2019) Green inertial sensors based on bacterial cellulose. In: 2019 IEEE sensors applications symposium (SAS), Sophia Antipolis, France, 2019, pp 1–4. https://doi.org/10.1109/SAS.2019.8706112
69. Evans BR, O'Neill HM, Malyvanh VP, Lee I, Woodward J (2003) Palladium bacterial cellulose membranes for fuel cells. Biosens Bioelectron 18(7):917–923
70. Torresa Fernando G, Troncoso Omar P, Gonzales Karen N, Sari Reka M, Gea Saharman (2020) Bacterial cellulose based biosensors. Med Dev Sens. https://doi.org/10.1002/MDS3.10102
71. Liang Y, He P, Ma Y, Zhou Y, Pei C, Li X (2009) A novel bacterial cellulose-based carbon paste electrode and its polyoxometalate-modified properties. Electrochem Commun 11:1018–1021
72. Yamanaka S, Watanabe K (1994) Applications of bacterial cellulose in cellulosic polymers. In: Gilbert R (ed) Hanser Publishers Inc, Cincinnati
73. Campano Cristina, Merayo Noemi, Negro Carlos, Blanco Angeles (2018) In situ production of bacterial cellulose to economically improve recycled paper, properties. Biomac. https://doi.org/10.1016/j.ijbiomac.2018.06.201
74. Basta AH, El-Saied H (2009) Performance of improved bacterial cellulose application in the production of functional paper. J Appl Microbiol 107:2098–2107
75. Zhang Mingquan, Xiao Wu, Zhenhua Hu, Xiang Zhouyang, Song Tao, Fachuang Lu (2019) A highly efficient and durable fluorescent paper produced from bacterial cellulose/eu complex and cellulosic fibers. Nanomaterials 9:1322. https://doi.org/10.3390/nano9091322
76. Yang J, Zhao C, Jiang Y, Han W (2016) The research of adding bacterial cellulose to improve the strength of long-fiber paper. In: 4th international conference on machinery, materials and computing technology (ICMMCT 2016), Atlantis Press

Printed in the United States
by Baker & Taylor Publisher Services